SONIC REBELLIONS

Sonic Rebellions combines theory and practice to consider contemporary uses of sound in the context of politics, philosophy, and protest, by exploring the relationship between sound and social justice, with particular attention to sonic methodologies not necessarily conceptualised or practiced in traditional understandings of activism.

An edited collection written by artists, academics, and activists, many of the authors have multidimensional experiences as practitioners themselves, and readers will benefit from never-before published doctoral and community projects, and innovative, audio-based interpretations of social issues today. Chapters cover the use of soundscapes, rap, theatre, social media, protest, and song, in application to contemporary socio-political issues, such as gentrification, neoliberalism, criminalisation, democracy, and migrant rights. *Sonic Rebellions* looks to encourage readers to become, or consider how they are, Sonic Rebels themselves, by developing their own practices and reflections in tandem to continue the conversation as to how sound permeates our socio-political lives.

This is an essential resource for those interested in how sound can change the world, including undergraduates and postgraduates from across the social sciences and humanities, scholars and instructors of sound studies and sound production, as well as activists, artists, and community organisers.

Wanda Canton is an artist and researcher currently completing her doctorate at the University of Brighton, UK.

SONIC REBELLIONS

Sound and Social Justice

Edited by Wanda Canton

LONDON AND NEW YORK

Designed cover image: Geoff Lewis and Wanda Canton

First published 2024
by Routledge
4 Park Square, Milton Park, Abingdon, Oxon OX14 4RN

and by Routledge
605 Third Avenue, New York, NY 10158

Routledge is an imprint of the Taylor & Francis Group, an informa business

© 2024 selection and editorial matter, Wanda Canton; individual chapters, the contributors

The right of Wanda Canton to be identified as the author of the editorial material, and of the authors for their individual chapters, has been asserted in accordance with sections 77 and 78 of the Copyright, Designs and Patents Act 1988.

All rights reserved. No part of this book may be reprinted or reproduced or utilised in any form or by any electronic, mechanical, or other means, now known or hereafter invented, including photocopying and recording, or in any information storage or retrieval system, without permission in writing from the publishers.

Trademark notice: Product or corporate names may be trademarks or registered trademarks, and are used only for identification and explanation without intent to infringe.

British Library Cataloguing-in-Publication Data
A catalogue record for this book is available from the British Library

Library of Congress Cataloging-in-Publication Data
Names: Canton, Wanda, editor.
Title: Sonic rebellions : sound and social justice / edited by Wanda Canton.
Description: Abingdon, Oxon ; New York, NY : Routledge, 2024. |
Includes bibliographical references and index.
Identifiers: LCCN 2023054859 (print) | LCCN 2023054860 (ebook) |
ISBN 9781032420622 (paperback) | ISBN 9781032420660 (hardback) |
ISBN 9781003361046 (ebook)
Subjects: LCSH: Music–Political aspects. |
Social justice. | Protest movements.
Classification: LCC ML3916 .S6588 2024 (print) |
LCC ML3916 (ebook) | DDC 306.4/842–dc23/eng/20231226
LC record available at https://lccn.loc.gov/2023054859
LC ebook record available at https://lccn.loc.gov/2023054860

ISBN: 978-1-032-42066-0 (hbk)
ISBN: 978-1-032-42062-2 (pbk)
ISBN: 978-1-003-36104-6 (ebk)

DOI: 10.4324/9781003361046

Typeset in Sabon
by Newgen Publishing UK

CONTENTS

List of Contributors — vii

Introduction — 1
Wanda Canton

1 Listening to Gentrification: Finding Socially Just Ways to Listen to Our Environments Together — 14
Bethan Mathias Prosser

2 'Made in LDN': Young People's Production of Rap Music in the Neoliberal Youth Club — 45
Baljit Kaur

3 ~~Dangerous~~ Dada? Reconceptualising UK Drill as Avant-Garde — 64
Wanda Canton

4 Memetic Feminism and TikTok — 91
Kathryn Zacharek and Wanda Canton

5 Acousmatic Sound, Neoliberal Anxiety and Theatrical
   Resistance   112
   Hara Topa

6 'Remove Them All!': Sounds of Protest in the Algerian
   Hirak Movement   132
   Stephen Wilford

7 Border Spaces and Sounds of Resistance: Music at the
   Franco–British Border   152
   Celeste Cantor-Stephens

Acknowledgments   *180*
Index   *182*

CONTRIBUTORS

Wanda Canton is an artist and abolitionist researching the criminalisation of rap music and socio-cultural forms of policing. Her publications have explored rap for trauma recovery, decoloniality, and she is currently developing concepts of radical listening practices. She is a podcaster host/producer, and former mental health practitioner with an interest in audio arts, peer support and working with incarcerated people.

Celeste Cantor-Stephens is a storyteller: a musician, activist, educator-facilitator, and researcher-writer, working at an intersection of arts and social justice. Across her work, she uses creative, interdisciplinary approaches as an expressive, human-led means of exploring and confronting. She holds an MPhil (University of Cambridge) and MSt (University of Oxford).

Baljit Kaur completed her PhD at the University of Sussex. Her research provides an ethnographic account of young people's experiences of violence in East London, and the ways in which they are narrated through the production of rap music. She currently teaches on the themes of rap and resistance.

Bethan Mathias Prosser's PhD developed a form of participatory listening methods to investigate residential experiences of urban seaside gentrification and displacement injustices on the UK south coast. Alongside research, Bethan works with interactive listening walks, sound foraging and music-making activities. She is also a violinist and performs in acoustic folk-punk bands.

Hara Topa trained in Dramatic Arts in Thessaloniki, Greece. She has acted in a number of theatrical productions and codirected with playwright Despoina Kalaitzidou. In 2016, she acted in Morteza Jafari's film *Isis Bride*. She has been working as an English and Drama teacher since 2000 and is currently a postgraduate researcher at the Royal Central School of Speech and Drama.

Stephen Wilford is Assistant Professor of Ethnomusicology, Popular Music and Sound Studies in the Faculty of Music at the University of Cambridge, and a Fellow of Wolfson College, Cambridge. His work focuses upon the musics and soundscapes of Algeria, in historical and contemporary contexts. He is Treasurer of the British Forum for Ethnomusicology and a member of the Royal Anthropological Institute's Ethnomusicology-Ethnochoreology Committee.

Kathryn Zacharek is a postgraduate researcher working on the biopolitics of right-wing populism and its intersection with race. Zacharek has previously worked as the assistant editor for the student-led journal Interfere titled *Disrupting Coloniality, Recovering Decoloniality?* She currently works as an Academic Support Worker for the University of Brighton and is an aspiring journalist.

INTRODUCTION

Wanda Canton

In 2020, the world fell to a quiet standstill. COVID-19 left streets and neighbourhoods locked down in silence, interspersed with conspicuous coughs and the sighing iteration: 'you're on mute.' Relieved by momentary ruptures of doorstep clapping for keyworkers and balcony concerts, our ordinary lives so full of sound became morbidly, sonically bereft. Our mouths muffled beneath masks, loved ones were comforted through windows, our speech dislocated through the partition of Zoom calls. We began to wonder if life would ever be the same. In May of that year, George Floyd was murdered by police officer, Derek Chauvin in Minneapolis, US. Suddenly, the world began to shout and cry, his final words carried over the voices of thousands of people and the Black Lives Matter movement, 'we can't breathe.' Cities and towns were engulfed by protests and unrest as people poured into the vacant streets, demanding an end to deaths in police custody. Less than two years later, Chris Kaba, a Black 24-year-old rapper and soon-to-be father, was shot dead by British police on the streets of London. The sonic paradox of 2020 had blatant acoustic dimensions but its conflicts of quiet/noise were by no means unique, albeit extreme. As the sounds of Kaba's family's grief pierced across Parliament Square, the crowd stood in solemn, unmoving silence.

'Sound is movement' (LaBelle, 2018: 95), which passes not only between human subjects, but transcends geographical, political, and cultural borders, adapting to localised contexts. In 2021, Indian farmers led a noisy tractor convoy parade to protest Prime Minister Modi's agricultural 'reforms,' which would see small farmers ostracised from a monopolised market. In Iran, 2022, following the killing of Mahsa Amini in police custody after she was detained for wearing an 'improper' hijab, women and girls led protests in the streets, amplifying their collective voice. Hundreds have lost their lives during the protests, with 14,000 people detained in the first few months, and some, including rappers and active social media users, threatened with the death penalty (Parent and Habibiazad, 2022), the ultimate attempt to silence dissent. The UK trialled its new emergency text alert, sending a siren-like message to all mobile phones in 2023 with many comparing it to the 'four-minute warning' system conceived in the Cold War, warning of imminent nuclear attack. Explosive bombs have continued to drop across Ukraine and recently again in Palestine. Violence, and its sonic dimensions, is a global emergency. Our changing acoustic landscapes are both natural (non-human) and man-made, with the crushing floods and blazing fires of climate change giving sound to our struggling environment. Two-thirds of bird species in North America face extinction including the wood thrush, a well-known songbird (Holden, 2019). The symbolism of this threatened songbird is representative of this book's intention: in locating the suppression of sound or acoustic oppression, there are pockets of resistance that can be found and, ultimately, hope for a more just world. One in which the songbird, metaphorical or otherwise, lives on.

Sonic Rebellions asks, 'what is the relationship between sound and social justice?' Two common themes emerge throughout this book, discussed further in the following sections. Firstly, we champion the rebellions of those who may not meet the ideal activist type, nevertheless finding agency and voice through their localised sonic practices. For example, rappers and social media users or neighbourhood walks and football stadiums may not be typically expected to facilitate political discussion yet become key players and sites in demanding change. We also recognise the contributions beyond the practical making of sound; *listening* repeatedly emerges as a key intervention. These actors, sound-makers, and listeners are shaping and reflecting physical as much as sonic architectures, particularly as it pertains to public space. Sound then, has a spatial capacity. Within the context of social justice, this can be a tool with which existing borders and exclusions can be challenged; whether public appearance and commentary, freedom of movement, the online public, and sites of political possibility. These spaces may not have been fully realised yet. Several chapters were written whilst the themes and practices they are concerned with developed in real-time, thus they respond to ongoing debates.

If listening has the potential to be an intervention, perhaps reading does too. As space is conceptual before tangible, the imagination of you, reader, is valued as interpreting the ideas within this book and developing them in your own practices and fields. The authors are artists, academics, and/or activists. They speak to a range of disciplines across the humanities, arts, and social sciences, whilst many apply their own lived experience and personal reflections. The book is international in scope, with chapters spanning across the UK, US, Greece, France, and Algeria. Within these localities, we pay particular attention to the divisions of racialised, gendered, or class communities, and the complex repercussions of neoliberalism.

In May 2022, *Sonic Rebellions* held its inaugural two-day symposium at the University of Brighton (UK), supported by the Centre for Applied Philosophy, Politics, and Ethics (CAPPE) alongside the Brighton Fringe Festival. Founded by Wanda Canton, the event was conceived as an opportunity to innovate the traditional conference style of spoken papers. Delegates were encouraged to utilise creative approaches in their delivery, with a preference for experiential or interactive facilitation. Workshops provided a space for participants to explore sound-walking, collaging, mixtape making, and listen to audio installations or seminars delivered through music. Genres included reggae, hip-hop, Jungle,[1] UK Drill, Grime, electronic dance music and djembe drumming. Emphasis was consistent across these varied mediums as to their accessible and inclusive methods, highlighting the capacity for sonic methods to overcome attempts to suppress and obstruct the connections between people. Keynote speaker, rapper and activist, Kareem 'Lowkey' Dennis concluded the conference with the take-away message: 'resistance is fertile, not futile.'

This book features a selection of the contributors to the May conference either as authors or peer reviewers. They represent a wide range of research disciplines, and many chapters are written by early career or independent researchers, who often face obstacles and disadvantages in publishing. Some chapters are inaugural publications of research projects, including doctoral work, whilst others are developed by creative practitioners. The process of writing this book was designed to reflect the ethos of the project. Over the initial months, monthly online sessions were held for everyone to hold space together for half a day, sharing ideas or writing together. The spirit of collaboration and interdisciplinary thinking was evident at the conference, so offered a foundation with which the isolation typically experienced in the writing process might be mitigated. It was also intended as an opportunity to develop camaraderie for the duration of the project. Throughout the journey there were personal losses, theses submitted, new projects started, and others finished. Some balanced music tours, caring responsibilities, health challenges, multiple jobs, and/or teaching. Each chapter will have had its own individual journey, a culmination of learning, experimentation, and challenges, with an invitation that it may play a part in yours.

This book was also written at a time of national crisis for UK Universities, with rolling strikes and boycotts impacting contributors across the board both as authors and reviewers. At the University of Brighton alone, where *Sonic Rebellions* was founded, mass redundancies were announced in May 2023 leading to catastrophic levels of stress and concern for the present and future viability of the University and its community. The use of sound has

FIGURE 0.1   Conference poster, designed by Wanda Canton.

been central to opposing the funding cuts, be it defiant voices in meetings with senior management, to rolling, chanting, singing protests, and collaborative discussions on picket lines. It has also meant that many contributors across the board in different institutions were working under the enormous financial pressure of 100% pay deductions for national action short of a strike, uncertainty about imminent employment, and mounting workloads. Indefinite industrial action has left vibrant, bustling campuses eerily quiet. Leaked correspondence between university management teams has revealed an explicit desire to be punitive towards unionised staff, with some stating 'I'd prefer pain along the way' in reference to massive pay deductions (Meighan, 2023). Other management teams have favoured absolute silence, as though there is not an addressable other requiring a response. Hence the term 'deafening silence,' the lack of sound can be as penetrating as the disregarded noise of protest and community. The continued commitment of authors and reviewers to this book demonstrates the passion and dedication of many researchers, who are forced to navigate high levels of productivity with reduced resources and often without pay. The current landscape for higher education has also highlighted an ideological threat against research in humanities, social sciences, and the arts, which are disproportionately facing closures of courses, research centres, and lectureships. This includes many of the disciplines central to this book. These research areas do not always align with neoliberal expectations of productivity and are more likely to criticise inequities than support budgetary or governmental agendas. Indeed, Conservative Prime Minister Rishi Sunak declared his intention to 'get far tougher' on universities which 'are not producing the goods' and 'full of, you know, people who don't vote for us anyway' (Siddle, 2023). The dismantling of research opportunities and working/living conditions for researchers poses a very real threat to the disciplines explored throughout this text and the resources available for projects like *Sonic Rebellions*.

Whilst there is criticism of academia as self-superior or too distant from the 'real world,' *Sonic Rebellions* consistently considers empirical application. In other words, whether, or how, conceptual theory can enable us to better support and understand social movements and change. Given the professional and personal practices of many of the authors and conference participants, some will have been inspired by the sonic medium, or actions for justice; turning to theory secondarily. Feminist theorist, bell hooks (1991) acknowledges how personal narrative can contextualise the motivations for theory and analysis. She writes of theory as a mode of healing; from the pain and suffering of injustice, which is given a voice through conceptual discussion. For hooks, theory is a social practice. She is clear that neither theory nor academia is empowering by default and can be used to create hierarchy or divisions of prestige. This includes gatekeeping as to standards of intellectual discussion, overly convoluted and jargonistic language, and narrowly traditional thematic considerations. This project has sought to challenge what is discussed within academia, centring contemporary practices from music to social media, and to widen the scope of participation through the use of creative styles and digestible language. The authors were encouraged to approach their text creatively, inspired by Katherine McKittrick's (2021) decolonising text, not only to embrace non-literary forms such as music, but to consider how these experiences are shared. Whilst some chapters take a more established or traditional academic format, others interweave a range of materials, including personal narrative, images, scores, and interactive suggestions to experience. Although this work is a written excursion in the *Sonic Rebellions* project, readers are invited to 'feel-with,' to use McKittrick's term (2021: 70), and to feel involved in sharing ideas. This at times includes addressing you directly, as an offer to extend and continue the conversations herein.

Sonic: Audio Non-activism

Activism is the *intentional* use of doing or not doing, making or not making something, which changes the world, environment, or behaviours in some way (Barber, 1984). It could be a campaign, event, or performance (in its broadest terms). For example, Environment Activism may include the intentional use of direct action to advocate for environmental protections. As this book developed, there were discussions as to whether it advocates a framework of 'audio activism.' In light of the earlier definition, audio *activism* suggests the deliberate use of sound-based practices for political purposes. However, such a term would contradict the first key finding of this work: that non-political agency is still political, that change is not always conscious or intentional. This does not seek to challenge the proactive and deliberate nature of activism, rather to expand the scope of actors effecting socio-political change. Authors recognise the significance of practices by those not commonly perceived to be activists and the spaces they occupy. For example, Prosser considers the acoustic observations of local citizens, Canton proposes young contemporary rappers as producing the new avant-garde, Zacharek and Canton explore the use of social media by young women and girls, Topa considers the affective reactions of theatre audiences. Wilford notes the political significance of Algeria's football stadiums and Cantor-Stephens reflects on moments of musical solidarity in the Franco–British border camps.

Secondly, an action may be centred on sound without this being informed by specifically sonic politics. For example, the company Shell's annual general meeting was disrupted in May 2023 by protestors interrupting speakers and collectively singing a protest song; 'go to hell, Shell' to the tune of Percy Mayfield's *Hit the Road, Jack* (Jolly, 2023). The action was clearly intended to intervene and used noise or song to do so, but this was just one strategy in a varied and creative movement, which utilises a number of other tactics, which may or may not use sound. To focus only on the sonic aspects or audio activism may risk dismissing the greater context and could make assumptions of the protesters' decisions.

Another interpretation of audio activism could be the use of political action to elevate acoustic methodologies/issues. This could, for example, concern politics of discourse and speech, whereby attention is called to those who dominant or are absent from acoustic spheres. Fields such as decoloniality critique privileged noisemakers and regulators but do not reduce their scope to consider only how colonial mentalities operate at a linguistic or sonic level. The eclectic forms of sound explored throughout this work are demonstrative of LaBelle's (2018) *sonic agency* within which 'individuals and collectivities creatively negotiate systems of domination, gaining momentum and guidance through listening and being heard, sounding and unsounding particular acoustics of assembly and resistance' (p4). *Sonic Rebellions* considers how socio-political subjectivity is enacted and expressed through sound. However, the interdisciplinary focus may mean that research designations could be better described as Audio Criminology, Acoustic Politics, Sonic Philosophy, and so on. This would maintain the primacy of the existing discipline, whilst demonstrating an exploration through sound, which enables the acoustic to be considered at both an ontological and epistemological level.

Feld (2015) describes *acoustemology* as the meeting of acoustics and epistemology; what is known through sound, both in producing and listening to it. The latter reveals a sonic agent who plays as crucial a role: listeners. Projects like the international Listening Biennial, founded in 2021, draw attention to listening as an intervention. The work of its artistic director, Brandon LaBelle, is referenced throughout this book and has been instrumental in conceptualising sonic agency. As philosophical cliches have queried (if a tree falls…), sound is a bilateral interaction (at least), for the speaker, musician, sound-creator relies on being heard, and this listening may be a key political act itself (see Prosser on soundscapes or Canton on UK Drill). That is, sound as a communicative act suggests a site of mediation or struggle between at least two subjects. Listeners may be critics or fans of what they hear, and they may or may not participate in the sound directly. This contributes to listening's capacity to be democratic and equitable, as politics is not simply finding common interests but navigating conflict and difference (Bickford, 1996). Perhaps, like the diversity of sound-creators discussed throughout this book, listeners can include readers, who may be limited in responding sonically due to its written form, but nonetheless engage and respond. Many of the chapters herein offer exercises for the reader to practice; or draw upon popular media or technologies with which readers might consider further engaging with or making themselves.

Sound is increasingly considered in participatory research as a method with which to engage the communities it concerns (Woodland and Vachon, 2023). This is reflected in the methodological decisions of authors such as Prosser and Cantor-Stephens, whose use of music, or listening, is central to developing rapport and insight from the communities they centre. Other chapters do not prioritise participatory research to create new data, rather listen to what already exists (see Canton; Zacharek and Canton; Wilford). This may be a conscious decision not to duplicate the sounds a community already makes on their own terms and time. Although research is increasingly centred on participation, often with good reason, such as avoiding the ivory tower of academia, some projects may ethically seek to avoid superseding how communities already engage with sound (sonic epistemes). Indeed, communities should have the right *not* to participate, without compromising being heard. The use of participatory methods in some cases may, if inadvertently, disregard an existing sound practice and elevate the researcher's idea of what something should sound like. This is not to suggest that participatory methods are colonial, indeed the communities being discussed should have a right of reply to research. However, as participation is increasingly a requisite for commissioned projects, there is a risk of colonising sound practices by extracting them from their organic use to meet the priorities or language of academic research. As Feld (2015) notes, acoustemology should neither be sonocentric. This suggests that the way in which knowledge is produced and experienced should consider other factors at play, beyond the acoustic. For example, Prosser emphasises the significance of diversity in listening experiences and how listening may be impacted by or intersect with other sensory forms (like sight and touch) and power dynamics, including colonial contexts. Again, this suggests that a label of audio activism may be too limited for the scope of the projects considered here.

Rebellions: Sonic Justice

Sonic Rebellions questions how sound is used to further or obstruct justice. Social justice is not easy to define and may be contested within any given school of thought including psychology, education, social policy, and philosophy, among others (Shriberg et al., 2008; Reisch, 2002). The subjective nature of justice is evident, if you will, in issues of criminal justice wherein some advocate for carceral practices to uphold justice, and others deem criminalisation and incarceration as an injustice in of itself. However, there are some shared sentiments in regards the authors' conceptualisations of justice. While some are explicitly concerned with neoliberalism, others critique the infrastructures that uphold it. As Zacharek and Canton explain, by way of Foucault, there is no central institution of power; rather it is dispersed and reiterated through a range of discursive and social methods which entrench specific expressions of power and dominance. There is a consistent critique throughout this work of policing: as an institution, of migration, of public and communal space, and cultural practices. Policing is conceptualised as a tool of cultural regulation and control rather than solely a profession, enacted by state and individual actors. This work finds therefore, that neither power nor resistance is limited to institutional behaviours; the policing of culture, power and rebelling for change can be realised across micro/macro levels.

Given our centrality of sound, our interests may be better described as *sonic justice*, which recognises both how life is organised by sonic hegemonies but also how it can be changed. Social justice is frequently conceptualised throughout this book as concerning space, or in LaBelle's (2018) terms, *unlikely publics*, which operate within day-to-day relationships rather than mass movements. Many of the practices explored throughout can be individually exercised, and do not require political or technological specialism. Chapters 1 through 5 consider practices and music forms in their capacity to be easily accessible or available by creators or audiences. This includes forms of listening emphasised as an embodied experience. They also refer to readily available technology including social media and community resources.

However, there is consideration of the strength of collective action, which recognises how localised events can impact a broader landscape by amplifying sound to challenge the exclusivity and oppressive regulation of public space. Such space is vital to conceptualising the configurations within which sonic agents are produced and sustained. That is, an acoustic ecology regulates or allows different subjectivities to exist or be heard (LaBelle, 2021). This is notably significant in cases where certain groups or demographics refuse to be silenced by existing policies and practices, which serve to prevent or obstruct them from public space. Chapters 6 and 7 consider the use of culturally and context specific music to advocate for civil and human rights. They also demonstrate the potential of music to build international solidarity. Katherine McKittrick, who has long inspired the *Sonic Rebellions* project, explains:

> Musical subversion is, importantly, tied to the development and legitimation of new modes of social kinship relations, reciprocal exchanges that do not replicate colonial heterosexual family figurations or individualist models but, rather, establish networks that collectively rebel.
>
> *(McKittrick, 2021: 163)*

Music is certainly a key interest throughout this book, although it is argued that other sonic methods are equally valuable.

Chapter Overview

The chapters in this book can be read as per the interests of the reader, in isolation or in conversation. It is hoped that attention to one methodology, theme, or context will pique curiosity to explore the remaining chapters which perhaps touch on something you may not have considered before. In parts, authors will refer to other chapters of relevance throughout the book, demonstrating that across different specialisms, we explore and think in collaboration or dialogue, noting points of alignment with authors of other disciplines.

Prosser's opening chapter offers listening exercises addressed to and *with* the reader. These reflective practices may be incorporated or used to inform a foundation with which you can engage with the remaining chapters. Through her experimentation of participatory and reflective listening, including in the context of a written chapter, Prosser invites readers to directly experience her doctoral work *listening-with* (Prosser's term), in the context of gentrification. For example, she notes the sounds of scaffolding and construction as acoustic signifiers of the changing seaside towns of the UK south coast, more acutely pronounced following the silence of the 2020 pandemic lockdowns. Building on Dylan Robinson's (2020) critique of 'hungry listening,' Prosser champions a decolonising approach to listening which, whilst acknowledging the relative privilege of her participants, prioritises listening diversity and choice in the production and sharing of knowledge. Prosser's reflections upon the class and racial dimensions of public space aligns with a concurring theme throughout this book as to how acoustic architectures reflect socio-political inequalities.

Kaur also reflects upon her doctoral work, at a youth club in East London. She explores the production of rap music, particularly Grime, in the context of austerity policies and according to the commercial opportunities provided by community services. Kaur supports the description of rappers as organic intellectuals, who represent a socio-political underclass even if their lyrical content is less obviously so. However, examining the way in which public funding influences rap cultures, she notes how youth centres central to artistic development have been subject to neoliberal agendas and financial cuts, disproportionately impacting marginalised young people who rely on these services for access to technology and music production. Subsequently, the priorities of youth centres shifted to crime prevention and rap programmes were co-opted into behaviour management. Exploring a particular youth centre's record label, Kaur argues that young rappers engaged with music programmes and increasingly subject to entrepreneurial ambitions and commercial productivity rather than self-expression as a value in of itself. 'Keeping it real,' in popular rap parlance, has diverted to 'making it.' Kaur therefore departs from the general consideration of how the sonic turns rebellious, instead problematising how forms which may have initially emerged as subversive, are increasingly commodified or adapted into neoliberal frameworks.

Contributing to this book's interest in rap music, Canton turns from Grime to UK Drill. By reconceptualising the subgenre as an avant-garde intervention, she critiques the ongoing criminalisation and censorship of the music and its artists. Drawing upon the use of every-day materials, a violence within and to language, and the aesthetic of masks and balaclavas, she compares UK Drill to the post-war Dada art movement. However, as a predominantly Black medium, racialised artists are not afforded the same artistic licence as their former, White Dada counterparts. The creative decisions of rappers are socially perceived as literal, rather than metaphorically or symbolically violent. The Black avant-garde is therefore deemed dangerous and threatening, accused of causing, rather than representing or mirroring social and community conflict, as Canton argues. She outlines the political and policing context within which UK Drill emerged, substituting the literal war protested by Dadaists for the historic and ongoing struggle against racism and Police[2] brutality.

Completing the book's consideration of accessible technologies, Zacharek and Canton propose that the social media app, TikTok, has played host to *memetic feminism* in their words. Following the recent surge of so-called 'alpha male' podcasters and rhetoric, women and girls have been quick to parody them on the same platforms misogynists seek to dominate. The authors highlight how misogynist speech via podcasts such as *Fresh & Fit* and *Tate Speech* are not reserved to the fringes of the internet but lead to serious ramifications, including allegations of human trafficking and abuse. They cite Foucault to explain how social media is used to promote patriarchal attitudes, by repeating and consolidating discursive practices which aim to naturalise and legitimise hegemonic masculinity. However, women and girls on TikTok have used mocking spoofs, reappropriated the speech of alpha podcasters, and utilised facial filters which draw comparisons to drag. These humorous memes ridicule the alphas, and, by evoking laughter and joy, they disarm the violence and monopoly of misogynist ideology and social media platforms.

Topa turns to the use of sound in theatrical productions. More specifically, the acousmatic: sound whose source is not seen. This, she argues, is a deliberate strategy of neoliberal governance to induce a continuous state of anxiety. She notes the European debt crisis, drawing attention to the perpetration of economic devastation by faceless institutions, which obscures accountability. A 2010 Greek production, *Cinemascope*, stages an apocalyptic Athens, demanding its audience use headphones and listen to the narration of a hidden voice. Topa draws on film theorist and composer, Michel Chion to demonstrate how this acousmatic voice is used to instil knowledge, control, and hierarchy. In total crisis however, like *Cinemascope*'s portrayal of the end of the world, power becomes futile. Topa refers to an impasse or gap, exposing the ultimate limits of authority and within which modes of resistance can manifest. She explores the use of ventriloquism as demonstrating how acousmatic sound can turn on itself, through dislocation and deadness, subverting vocal or corporeal limits and subsequently forming theatrical methods of resistance: the staging of survival.

Wilford outlines the *Hirak* ('movement') protests in Algeria, with 800,000 people marching through Algiers in February 2019 against Bouteflika's regime. He opens by recalling the interjection of Torki, a young man who contradicted a reporting journalist, proclaiming 'remove them all!' Wilford highlights Torki's use of local *Darija* Arabic as a linguistic rebellion, his demand echoing across broadcasts and social media, inducing debate as to Algerian rights. Mapping Algeria's anti-colonial and post-colonial periods,

Wilford highlights how musicians and poets have taken a leading role in resistance movements but have subsequently been targets of violence including the murder of outspoken musicians throughout the 1990s. This threat to life has moved the contemporary, anti-authoritarian *Hirak* protests into Algeria's football stadiums, whereby the crowd protects the identification of individuals. Whilst several chapters herein explore the creation of public space as bolstering visibility, Wilford demonstrates how public space can also be used to anonymise, at both visible and acoustic levels; providing an example of Topa's acousmatic rebellion. He contrasts this with Algerian Andalusi performances on international stages and explicitly political rap, which, whilst not concealing the identity of performers, acts as a staging of solidarity, subsequently working to unite North African audiences within and outside of Algeria.

The final chapter closes with Cantor-Stephens' interweaving of personal narrative and text scores for readers. Like Prosser's interactive style, she invites readers to perform or reflect upon migrant rights and borders by following the scores. Discussing her work as a music practitioner, Cantor-Stephens *underscores* (to emphasise and to provide actual scores) the role of music in broadening the restrictive space of the Franco–British border. Reflecting on her observations between 2015 and 2023 in the Calais refugee camps, she argues that borders, which extend beyond simple or coherent physical boundaries, create a *Certain Human's Land*, as she names it. That is, they are states or spaces of exception via Agamben, where rights are unevenly distributed by states and authorities. Music becomes a central vehicle for solidarity and humanity in these spaces. She highlights the significance of spontaneous gatherings, loosely comparing its ability to connect, protest and offer escapism, to reggae sound systems. She recites lyrics, both political or narrative in nature, by residents of the refugee camps, to elevate their stories, noting how music contributes to the recognition of individuals and peoples, particularly where they are otherwise dehumanised by oppressive policing and immigration practices. She finishes with an invitation to the reader to reflect upon their own agency and listening practices, which well concludes this book.

Conclusion

Sonic Rebellions conceptualises the use of sound by explicitly and non-explicitly political actors, which shapes the sonic architectures used to regulate and police social life. Social justice (or perhaps *sonic* justice) is conceived as a collective, spatial action which opposes neoliberalism and its infrastructural arms. This book intends to develop and widen the discussions at our inaugural conference, with an invitation for you, reader, to participate.

References

Barber, B. (1984) Strong Democracy: Politics as a Way of Living. In B. Barber. *Strong Democracy: Participatory Politics for a New Age*. California and London: University of California Press.

Bickford, S. (1996) *The Dissonance of Democracy: Listening, Conflict, and Citizenship*. Ithaca and London: Cornell University Press.

Feld, S. (2015) Acoustemology. In D. Novak and M. Sakakeeny (eds) *Keywords in Sound*. Durham and London: Duke University Press. pp. 12–21.

Holden, E. (2019) Two-thirds of bird species in North America could vanish in climate crisis. *Guardian*, 10 Oct, available at: www.theguardian.com/environment/2019/oct/10/bird-species-extinction-north-america-climate-crisis. Accessed 21 July 2023.

hooks, b. (1991) Theory as liberatory practice. *Yale Journal of Law and Feminism*, 4(1): 1–12.

Jolly, J. (2023) Shell AGM disrupted by protests as investors reject new emissions targets. *Guardian*, 23 May, available at: www.theguardian.com/business/2023/may/23/shell-agm-protests-emissions-targets-oil-fossil-fuels. Accessed 9 Sep 2023.

LaBelle, B. (2018) *Sonic Agency: Sound and Emergent Forms of Resistance*. London: Goldsmiths Press.

LaBelle, B. (2021) *Acoustic Justice: Listening, Performativity, And the Work of Reorientation*. London: Bloomsbury.

McKittrick, K. (2021) *Dear Science and Other Stories*. Durham: Duke University Press.

Meighan, C. (2023) University chief wanted to inflict 'pain' on marking boycott lecturers. *STV News*, 5 July, available at: https://news.stv.tv/north/aberdeen-university-chief-wanted-to-inflict-pain-on-marking-boycott-lecturers. Accessed 13 July 2023.

Parent, D. and Habibiazad, G. (2022) Rapper who protested over death of Mahsa Amini faces execution in Iran. *Guardian*, 11 Nov, available at: www.theguardian.com/global-development/2022/nov/11/rapper-who-protested-over-death-of-mahsa-amini-faces-execution-in-iran. Accessed 19 June 2023.

Reisch, M. (2002) Defining social justice in a socially unjust world. *Families in Society: The Journal of Contemporary Human Services*, 83: 343–354.

Shriberg, D., Bonner, M., Sarr, B.J., Walker, A.M., Hyland, M., and Chester, C. (2008) Social justice through a school psychology lens: definition and applications. *School Psychology Review*, 37: 453–468.

Siddle, J. (2023) Rishi Sunak vowed to get tough on universities as they're full of non-Tory voters. *Mirror*, 1 July, available at: www.mirror.co.uk/news/politics/rishi-sunak-told-meeting-hed-30370275. Accessed 13 July 2023.

Woodland, S. and Vachon, W. (2023) *Sonic Engagement: The Ethics and Aesthetics of Community Engaged Audio Practice*. Oxon: Routledge.

Notes

1 Throughout this work, capitals will be given to some genres of music to distinguish it as a proper noun (Jungle) rather than the common noun (jungle).
2 A capitalised Police refers to the institution as opposed to the verb to police.

1
LISTENING TO GENTRIFICATION

Finding Socially Just Ways to Listen to
Our Environments Together

Bethan Mathias Prosser

> Will you listen with me?
> How do we listen together?

This chapter starts with an invitation to you, the reader. I invite you to listen with me. To your surrounding environment. And specifically, to how gentrification might be resonating around you.

It is an ambitious ask, an experiment in participatory listening research. The overarching ambition is to find out how we can support and enable listening for the purposes of learning about, and potentially acting on, issues of social justice. Participatory listening research is a way of listening with others, to our environment, that generates new understandings whilst embracing different listening experiences, practices, and positionalities (Prosser, 2022). This chapter asks, how can participatory listening research be used through the medium of a book chapter?

This listening experiment focuses on gentrification and its associated injustices. Gentrification is a contested term coined over 60 years ago (Glass, 1964), which I consider contagious given it has since exploded into a diversified, globalised, and controversial phenomenon. It can be defined in critical geography terms as 'the production of space for progressively more affluent users' (Hackworth, 2002: 815). I draw on my PhD research into gentrification at the seaside, in which a participant revised this definition to 'the poshing up of a place to the detriment of working folk' (Dr X,[1] Interview 11). Myriad injustices are associated with gentrification processes, which stem predominantly through the displacement of existing 'working folk' due to this 'poshing up.' Forced movement is the most easily recognisable injustice, but, increasingly, displacement caused by gentrification is understood to encompass an expanded range of experiences that operate on a spectrum. This includes complex feelings of alienation and estrangement, the loss of social networks and accessible services, the effects of unequal housing and redevelopment and exclusions created by an increasingly dominant monoculture (Marcuse, 1985; Davidson, 2009; Atkinson, 2015; Elliott-Cooper et al., 2019; Phillips et al., 2021).

How can we tune into these exclusionary changes through listening in our homes, streets, neighbourhoods, towns, and cities? Through this research project, I have created a form of participatory listening research, called *listening-with*. The project used this listening approach to investigate urban seaside residents' experiences and understandings of gentrification on the UK south coast. Undertaken during fluctuating lockdowns in 2020, I was compelled to create a way of listening-with residents remotely. These methods included participants designing their own listening walks and listening-at-home activities, creatively capturing sounds, sensations, thoughts and reflections and sharing these through follow-up interviews. As the researcher, I was physically separate from participants and their neighbourhoods. I was reliant on digital and telephone technology to listen from a different location or listen back at a different time to recordings. Nevertheless, we found a way of listening together and reflecting on gentrifying changes and their effects. As an author, I am similarly separated from readers and their neighbourhoods, spatially and temporally. Thus, in principle, this approach of remote listening can be applied through written form. This is the starting premise of this chapter.

Striving to bring academia and practice together, I offer listening exercises and prompts for you to undertake whilst sharing findings about listening to gentrification. I first introduce different approaches to listening, mainly using the work of sound artist Pauline Oliveros (2005) and sound artist/academic Dylan Robinson (2020). Secondly, I give an overview of my research project and its methods. Thirdly, I offer two exercises for listening to gentrification and its injustices and discuss findings from listening-with research participants to *urban seaside gentrification*. I conclude by reflecting on the potential for finding more socially just ways to listen together. Ultimately, it is up to you – how you choose to engage, listen, and use this material. I do not know where you listen and what impacts and shapes your listening. I trust your judgement in choosing how you go about this safely for you. I hope to spark your sonic imagination and curiosity for how your changing environment resonates around and with you. And in the process, experiment with how we can listen together to issues of social justice.

16 Bethan Mathias Prosser

Listening practices

> What is sound?
> What is listening?
> What action(s) is usually synchronised with sound?
> When do you feel sound in your body?
> What sound fascinates you?
> What is a soundscape?
> What are you hearing right now? How is it changing?
> *(Nos. 10–16 from Listening Questions, Oliveros, 2005: 55)*

Sound studies encompass an expanding, dynamic, and varied range of approaches to the above Listening Questions about sound, listening, and soundscapes. Conventionally, a division is made between the physical characteristics of sound sources and how these characteristics are perceived and processed (Guillebaud, 2017). Although used interchangeably in common usage, there is a distinction made between 'listening' as an active process and 'hearing' as passive (ibid.). In this section, I provide an overview of key approaches to listening from sound studies. I offer this discussion as helpful in practice, for thinking about and revisiting these Listening Questions and those offered later in this chapter.

Early academic interests into the significance of listening to our environments are found in anthropology (Feld and Brenneis, 2004), ethnomusicology (Post, 2006), and sound art (Schaeffer et al., 2012; Drever, 2013). Sound scholars extol the qualities of sound in an attempt to push back against the dominant Western hierarchy of the senses. This hierarchy elevates the visual over all other senses, including within research methods (Howes, 2005). For example, this visual dominance is evident in gentrification research with few studies delving into sound to explore urban change (Guiu, 2017; Sánchez, 2017; Martin, 2021; Summers, 2021; Prosser, 2022). The soundscape movement laid several foundations for advancing sound and listening (Schafer, 1994). A soundscape is any acoustic environment that listeners experience surrounding them 'in space' (Schafer, 1994: 7; Helmreich, 2010: 10). By investigating soundscapes, acoustic ecology has created a comprehensive methodological toolkit with terminology, listening practices and soundwalks that academics and practitioners continue to utilise and advance (Bull, 2018: xxii). Many of these approaches include participatory aspects, such as equipping people with the tools for exploring and designing their own acoustic environments (Uimonen, 2011). For example, listening preference tests have been used with pupils to identify what would make their school environments more conducive for learning and subsequently demand changes from school administration (ibid.). Some approaches are also concerned with issues of social justice, such as how soundscapes can be interpreted as indicators of social conditions and societal trends (Arkette, 2004; Summers, 2021). Unlike other creative methods, such as photo voice (Wang, 2006), participatory video (Butcher and Dickens, 2016), and participatory mapping projects (Herlihy and Knapp, 2003), listening methods are not often explicitly coupled with or titled as a participatory research approach.

To develop participatory listening research, I sought out listening practitioners that recognise, value, and support that sound is experienced in a multitude of ways. Increasingly, there are critiques of approaches in sound studies, which do not account for such difference. Listening is frequently talked about as a universal, 'natural,' and transformative experience for all (Chaves and Aragão, 2021). But these 'typical hearing' assumptions create exclusions and miss out on the rich and significantly diverse ways people listen and relate to sound (Drever and Hugill, 2022). They do not always consider the difficulties, challenges, and uncomfortable experiences of listening, which came up in my seaside research (see next section discussion). Many scholars from Deaf,[2] queer, and decolonial studies assert the diversity of listening experiences and practices (Haualand, 2008; Bonenfant, 2010; Friedner and Helmreich, 2012; Harold, 2013; Drever, 2019; Robinson, 2020; Chaves and Aragão, 2021). These critiques are significant for thinking about listening with others using a participatory listening research approach. A participatory ethos embraces different ways of knowing, challenges power dynamics, and strives for person-centred accessibility and inclusivity (Beebeejaun et al., 2014; Hall and Tandon, 2017). For example, participatory listening research would seek to support and learn from the specific ways you may have answered the above Listening Questions. It may bring researcher(s) and participant(s) together to analyse the answers and understand what shaped them. It could then potentially explore how to improve the questions or find ways of investigating that might be more accessible, relevant, and responsible.

There are two sound art practitioners, in particular, that provide insightful concepts and tools for listening with others to our environments: Pauline Oliveros (2005) and Dylan Robinson (2020). I draw on their work as a way of remaining open and embracing the plurality of listening experiences. Oliveros is renowned in developing Deep Listening as a tool for music composition but also for therapeutic and wellbeing benefits. Oliveros (2005: xxii) separates out hearing as the physical means that enables perception and listening as giving attention to what is perceived both acoustically and psychologically. She argues more is known about the former with accompanying measurements and metrics but listening remains a mysterious process (Alarcón and Herrema, 2017). For example, Westerkamp (2017: 30), a pioneer of soundwalks, argues that we all listen differently and listening is 'never static and implies constant shifting and perceptual movement.'

Typologies of listening have been developed to pin down this mysteriousness (Guillebaud, 2017). A common categorisation is made within these models between mundane listening (everyday, natural, gathers information), causal listening (interpreting a message or determining the nature of sounds), and specialised/reduced listening (focusing on the traits of sound itself, acting on sound as a musician) (Guillebaud, 2017: 12). Acoustic ecology explores this mysteriousness in a different direction through developing methods and training for listening to the environment (Wrightson, 1999). Over the last 50 years, the work of the World Soundscape Project has been influential in shaping dominant understandings of listening to the environment through asserting the mediating role of sound and corresponding need to develop 'sonic competence' (ibid.; Yoganathan, 2022). However, these typologies and approaches risk imposing unnecessary hierarchies onto different ways of listening, separating people out into trained/higher and unaccredited/lower listeners (Chaves and Aragão, 2021). As Yoganathan (2022) critiques, contemporary acoustic ecology has yet to seriously question problematic ideas about race and social difference that have been found at their foundations. More inclusive approaches that examine how 'the subjective perception of aural environments is deeply influenced by the racial, class, sexual, gender, and dis/ability identifications of listeners' (ibid.: 464) sit on the margins of sound studies (Carmona, 2020; Gutierrez, 2019; Martin, 2021; Summers, 2021).

Oliveros (1974; 2005) is striking against this backdrop of going beyond rigid categories of listening in her practice-based work. As she explains:

> *Deep* coupled with *Listening* or *Deep Listening* for me is learning to expand the perception of sounds to include the whole space/time continuum of sound – encountering the vastness and complexities as much as possible ... Such expansion means that one is connected to the whole of the environment and beyond.
>
> *(2005: xxiii, original emphasis)*

Her listening practice is primarily focused on sound art composition. However, the notion of 'deep' chimes with research purposes in striving for new understandings and ways of knowing. Oliveros (2005) describes 'deep' as being about the 'edges beyond ordinary or habitual understandings' and defying 'stereotypical knowing.' She similarly talks about attention and connections to the external/global and internal/self (2005: xxv–xxiv). The Listening Questions at the beginning of this section demonstrate the type of questioning Deep Listening explores through a series of exercises to facilitate 'a community of creative interest' (ibid.: 57). Although Oliveros does not explicitly explore the different listening practices in the ways that recent critiques interrogate, her emphasis on openly facilitating creativity serves to not close down or exclude different ways people might listen.

Dylan Robinson (2020) develops an approach to listening that pushes further and gives a critical edge to understanding listening diversity. As an Indigenous Canadian sound scholar, he provides the most compelling recent work on decolonising listening in proposing a listening practice to remedy these critiques (Couture et al., 2020). Robinson's approach allows a continuum of listening practices and interrogates listening regimes that have been imposed and implemented (Robinson, 2020: 40). Arguing that our perception is acquired 'over time through ideological state apparatus,' he describes how Canadian settler institutions disallowed certain types of listening practices as part of their colonising endeavours (ibid.: 9). Robinson engages with sound studies from an Indigenous perspective 'to theorize and thematize listening as a political and cultural act' (Couture et al., 2020: 2). Robinson (2020: 40) develops resonant theory that challenges the Western focus on single senses. Within his theorisation, it is his concept of critical listening positionality that I take inspiration from in this project.

Positionality is a well-known academic concept, which explicitly considers the researcher's own position in the research process. It is often used to interrogate insider/outsider identities and power dynamics, such as the degree of lived experience the researcher shares with research participants or topic (Etherington, 2004). Robinson (2020) applies this to listening to understand how our individual and collective experiences, backgrounds, and ways of being in the world influence the way we listen. He argues that we carry listening privileges, biases, and abilities, not wholly negative or positive. But when we interrogate these through self-reflexive questioning, we start to engage with critical listening positionality:

> Like positionality itself, engaging in critical listening positionality involves self-reflexive questioning of how race, class, gender, sexuality, ability, and cultural background intersect and influence the way we are able to hear sound, music, and the world around us.
>
> *(Robinson, 2020: 10)*

Critical listening positionality is a powerful and challenging practice that brings political potency to the idea that 'we all listen differently' (Couture et al., 2020: 9). Robinson (2020) develops an ethics of listening to Indigenous music as an alternative to the extractive ways Indigenous content is predominantly consumed. He describes settler listening positionality as 'hungry listening' (2020: 2). This is grown from historical Indigenous encounters with settlers as 'starving people,' which referred to their bodily state and their hunger for gold. He shows how this hunger persists and infiltrates contemporary and dominant ways of listening.

You may not be reading and listening in a settler/Indigenous context, just as this research project has not taken place in this specific context. However, we are all connected to these listening regimes that shape listening, such as through the legacies of colonialism in our education systems. Most pertinently, 'hungry listening' lays bare the potential hunger of academia, which can be uncomfortable to grapple with:

> We in academia are locked in a race to stake our territorial claims, our quest for proof of originality the substance of our mimetic rivalry: to coin a term, to claim our corner of the knowledge market, to make something useful for the knowledge production of others while requiring acknowledgement of our role in that production.
> *(Randolph Jordan in Couture et al., 2020: 11)*

I attempt to respectfully engage with Robinson's concept, in the 'dialogical spirit' of his writing (Couture et al., 2020: 2) in developing participatory listening research. Robinson asks us to undertake challenging and detailed self-reflection that can enhance the ways we might be able to listen together and to social justice issues.

There are common threads running between Oliveros and Robinson's listening endeavours. Both strive to remain open to what listening can be. Through Deep Listening, Oliveros (2005: xxi–iii) attempts to defy 'stereotypical knowing' and simultaneously connect to our environment and 'consciousness':

> Consciousness is acting with awareness, presence and memory. What is learned is retrieved and retrievable. Information, knowledge and events, feelings and experiences can be brought forward from the past to the present. In this way one has self-recognition.
> *(ibid.: xxi)*

This ambition of consciousness-through-listening connects to Robinson's (2020) idea of critical listening positionality. He challenges us to undertake self-reflexive questioning about what shapes our listening whilst leaving this open to individual practice. In Robinson's (2020) concluding book discussion, he invites collaborators to reflect together on the possibility for this type of practice. He wonders, 'is it even possible to have one ear open to the unknown and another to one's positionality?' (2020: 249). Both Oliveros and Robinson are operating in the boundaries between sound art, listening practice, and academic thought. How can these concepts and practices be used in the realm of social research, and specifically, research that focuses on social justice and gentrification?

I argue in this chapter that participatory listening research can provide a way forward for such endeavours, embracing different listening practices to generate new understandings. I believe it is significant that Robinson (2020) chose a collaborative approach for his final reflections on critical listening positionality and other theorisations in his seminal book. Practicing with others is similarly at the heart of Oliveros' (2005: 57) development of Deep Listening. Both provide rich tools in finding ways to listen openly, critically, and reflectively with others. The challenge here is to use these through the medium of a chapter and to focus attention and self-reflexive questioning onto how gentrification is resonating around us.

The next section begins with an adapted Deep Listening exercise used in my gentrification research project with all participants. I invite you to listen wherever you might be, perhaps closing your eyes to help focus your attention before listening through these questions. These initial listening questions are offered as a way of tuning into your current environment. This is something we are not always accustomed to doing and can bring up a range of reactions from feeling calmer, pleasurable surprise to discomfort. Discomfort can bring new insights; as described by Dreher (2009: 451), listening can entail 'the recognition of not knowing as well as knowing ... unlearning as well as learning.' But, if you feel distress during these questions, consider pausing, stopping, or possibly focusing outwards rather than internally. I follow this exercise by explaining the method I created for listening remotely with residents to gentrification, before moving onto more specific exercises and discussion of gentrification-related listening.

Listening-with

I invite you to close your eyes for a few moments and then, on reopening, listen through the following questions:

1. First listening to the body: what internal sounds can you hear? Can you hear yourself breathing, your heartbeat?
2. Becoming aware of the sounds around you: what can you hear?
3. Are there any continuous sounds?
4. Are there any rhythms? Beats? Intermittent sounds?
5. What's the loudest sound you can hear? What's the quietest sound that you can just about make out?
6. What's the highest pitched sound you can hear? And what's the lowest pitched sound you can hear?
7. What else can you hear?
8. Are these sounds near to you?
9. Are they far away?
10. Can you identify where these sounds are coming from?
11. How do these sounds make you feel?
12. Do you like or dislike any of these sounds?

13. How do they affect your other senses? How do the sounds connect to what you can feel? What you might see? What you can taste? What you can smell?
14. What else can you hear?

(Adapted Listening Questions, Prosser, 2022)

I chose these Listening Questions to help research participants get used to listening to their surroundings, slow down, and stimulate different aspects of listening. The questions focus attention onto different types of sounds, qualities, and connections with other sensations and feelings. I share them with you, the reader, in a similar attempt, to help you start tuning into your surrounding acoustic environment. This preparation was crucial for initiating the participatory listening research approach that I developed throughout my PhD project. In this section, I discuss this approach in more detail. I give an overview of the methodology and pull out the key dynamics of this remote method. Discussion of participants' experiences, of being led through these questions and then undertaking listening activities in their neighbourhoods, may chime with your own experiences or spark further questioning. After discussing my specific approach, which I term *listening-with*, in the next section I invite you to undertake more specific listening activities where you are, to explore listening to gentrification.

Listening is the dominant practice that held my methodology together – explicit and central to every part of the project. There were four parts that all participants experienced in this method:

- A Deep Listening exercise led by the researcher remotely (as above).
- An immersive listening experience away from the researcher: listening walk or listening-at-home (20–30 mins).
- A way of capturing this experience whilst repeating the above activity detailed discussion with the researcher about the experience via phone or video call.

There were two main options that each participant chose: the type of listening activity and how they captured their observations about the experience. In addition, the participant chose where they listened: either a route taken around the neighbourhood as a listening walk or where in their home they listened. The participants also decided what technology to use: to capture their experiences and to communicate with the researcher (via phone or their choice of online software). Enabling participants to have these choices in the data production was part of the participatory ethos. But it also created the flexibility for conducting research during a pandemic, allowing activities to take place throughout varying lockdown states of restricted movement (see Brown and Kirk-Wade, 2021, for a breakdown of UK lockdown stages).

Whilst being responsive to government policy, it supported participants to make their own assessment of risk, allowing those shielding or uncomfortable with walking around the neighbourhood to participate from their home (as approved by the University of Brighton ethics committee).

Twenty-two residents living in three neighbourhoods across Brighton, Worthing, and St Leonards-on-Sea took part in these activities. As place-driven research, the only criterion was that the participants were living, at the time of the research activity, in one of the urban seaside neighbourhood sites that I designated. My participatory ethos and justice framework shaped my intentions to recruit people from a range of different social groups. However, the Covid-19 restrictions impacted recruitment (mainly via social media, professional networks with three targeted mailshots), resulting in a more privileged cohort of participants. Their ability to offer time to the research during this period of crisis speaks to this degree of privilege. For example, the majority of participants were owner occupiers, with four renters, one council tenant, and one in cooperative housing. Many were working in the public, charity, and arts sectors and the overwhelming majority had moved to the seaside as part of navigating the uneven affordability of the region. However, there was variation within this group, for example, two people had experienced homelessness. Taking a longer view over participants' life histories revealed how the majority had previously experienced the effects of gentrification-induced displacement in its broader sense.

Within a relatively small cohort, there was a range, with some limitations, of social characteristics. There was a balance of male and female participants (though no-one identified as non-binary or trans) and ages ranged from 20s to 70s. Predominantly, participants were White, with two participants identifying as Asian and two participants as mixed parentage. The majority also identified as heterosexual, with four identifying as lesbian, gay, or bisexual. These social identities and cultural backgrounds came to play a part in understanding how listening can generate knowledge about urban seaside gentrification (Prosser, 2022). Significantly for the method, several shared details of their disabilities: two used either a wheelchair or mobility scooter due to long-term illness and one person described being partially sighted. With regards to listening, two declared hearing impairments; one person with tinnitus and another with deafness in one ear. Although I am not able to go into comprehensive detail within this chapter, I invite the reader to think about the myriad factors that may shape your listening.

> How do your intersecting identity characteristics interact with your listening?
> How might your social positioning in your neighbourhood shape how you listening there?
> What privileges and marginalisations do you experience that might influence the way you listen?

24 Bethan Mathias Prosser

Overall, this method generated rich, nuanced, and multi-modal data that surpassed my tentative adaptations during heightened uncertainty in 2020. For capturing their experiences, participants rewalked the route or resat where they had been listening. I either called the participant who described their walk or sitting experience over the phone, which was audio recorded (option A). Or the participant chose to capture their observations independently using their own equipment e.g. audio recordings, photography, video, drawings, and/or notes (option B). The technology required to carry this out remotely was initially an unwelcome intrusion. Figure 1.1 shows the technology arrangement that relied on hardware on my side to be able to record the different ways participants chose to do the Deep Listening exercise, (option A) and follow-up interview.

I came to embrace the distinct research and listening practices that this technology set-up engendered. I used a creative and reflective listening practice to explore these dynamics. I created a sound collage from the 'behind-the-scenes' audio material recorded during the activities (for more detailed discussion of this piece, see Prosser, 2023). Through creating this sound piece, I became increasingly aware of how important the participatory aspects of the method were for generating new understandings. Participants

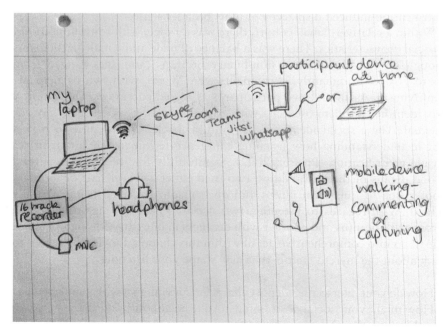

FIGURE 1.1 Drawing of researcher and participant technology. Drawing by the author.

had more choice over their listening experience, which produced a wider range of different material about gentrification. This method generated: 10 commented walk audio recordings, 17 participant recorded audio recordings, 235 participant photos, 53 participant video recordings, 22 pages of participant drawings, and 5 pages of participant notes.

Whilst this range of material created a challenge for analysis, it demonstrated all the different ways that participants were listening in the research. This was underpinned by the choice of participants' design. For the listening walks, the way participants chose their routes demonstrated the variety available: familiar walks that the participant undertakes regularly, new routes designed specifically to take in different sites and sounds, or roughly improvised wanderings through the neighbourhood. Only three participants chose to do the listening-at-home activities (one for mobility reasons and two due to shielding from Covid-19), but there was variety in their designs and capturing. One participant chose to listen and make audio notes at the front and then back of their flat. One participant wrote a page of notes on their roof terrace with accompanying photos. One participant wrote a series of 'sound minutes' at an open front window, noting sounds and observations every minute for an hour of listening. However, going beyond their choices in the method techniques, participants' listening experiences also differed in how sounds stimulated different responses, ideas, and reflections. I developed a technique of layered soundmapping to plot this different sound material, from which I identified a series of sound stimuli (Prosser, 2022). This focuses on how sounds stimulate emotional sensations and responses as well as trigger thoughts, discussions, stories, or memories.

> Thinking about your answers to the above listening questions that started this section, what sounds stood out to you? Were there any surprises? What did they trigger? Were there any specific ideas, memories, or stories?

These sound stimuli can be connected to different listening experiences. One sound stimuli was particularly prominent across all the listening-generated material, that of 'sound surprises.' Some listeners were surprised by what they could or could not hear, by the tension between their visual and aural sensations, and their unmet expectations and underlying sensory assumptions about their neighbourhoods. Some found listening transformative, whilst others found it difficult, struggling to engage with sounds over other stimuli, or conversely finding it too intense. 'Sound surprises' can be translated into thinking about certain forms of listening, such as 'unexpected listening,' 'difficult listening,' or 'transformative listening.' For example, one participant in Worthing, Rafael, surprised himself 20 minutes into his listening walk by suddenly singing out loud 'a non-sensical song about being ready for bed' (Listening Walk 03). He explained during the walk and the follow-up interview that singing was a reaction to the cumulative intensity of silent walking (Interview 03). It was triggered by a particular combination of the sounds of a wheelchair and smells of a bin lorry passing by at the same time. This is an example of a particularly expressive reaction to a 'sound surprise.'

A more common surprise across participants centred around the seaside. Many expressed surprised at how little they could hear the sounds of the sea and beach when walking nearby. Instead, many commented on the dominance of traffic along seafront roads and how the visual experience was more pleasant than the aural. Each resident brought their own identities and experiences to their sonic encounters, which shaped their responses. For seafront walking participants, their expectations were shaped by what they thought the seaside *should* sound like or how they imagined it to sound. This was challenged by their listening experiences in this project. In the above example, Rafael's usual way of walking around his neighbourhood influenced his reaction. Rafael's everyday walks along this familiar route would normally involve listening to music or taking his dog for a walk who he would talk to. But this experience was different. This method focused his attention onto all the sounds and sensations around him, which became increasingly intense the further he went into his journey. This example raises the ethical considerations of doing research remotely, bringing a different dimension to what it means to be an outsider/insider as a researcher (for further discussion of outsider/insider dimensions to remote research, see Prosser, 2023). It also strengthens the argument for acknowledging and embracing the different ways of listening in order to support these appropriately in research settings. For example, in my pre-walk discussion with participants I checked about any concerns or specific issues that might affect their walking and listening experiences. I asked them to keep mindful of pedestrian safety practices and I was on call for any incidents or issues arising.

Participants' listening experiences in this project chime with the ideas of consciousness-through-listening (Oliveros, 2005) and critical listening positionality (Robinson, 2020) introduced in the previous section. The majority of participants became aware of 'stimuli and reactions in the moment' (Oliveros, 2005: xxi). Listening to their neighbourhoods also brought up reflections and stories about past experiences which Oliveros believes brings us closer to 'self-recognition.' Furthermore, through developing a creative listening analysis approach, I was able to apply Robinson's (2020) idea of critical listening positionality for research purposes. I interrogated my own researcher listening positionality, driven by academic hunger and shaped by being unintentionally remote (Prosser, 2022). For example, my own 'typical hearing' positioning had previously led me to assume that people with any hearing impairments would not want to take part in the project. But two participants with hearing impairments challenged this. These participants helped me reflect and question these assumptions, drawing on ideas from aural diversity (Drever and Hugill, 2022). Participants could adapt the activities to suit their needs, which is part of the flexibility and person-centred aspects of a participatory approach. Understanding what was shaping participants' listening experiences became a way of analysing and interpreting the material. Through the interviews, I was able to discuss with participants why they had been stimulated by sounds in certain ways, which continued into thematic coding and the layered soundmapping. As will be highlighted in later discussions, this elicitation enabled the residents to delve into their thoughts, ideas, judgements, and emotional responses and begin the analysis process about gentrifying change in their neighbourhoods.

Listening-with is a way of referring to this specific analysis approach that I developed within this project. In hyphenating *listening* and *with*, I embrace the plural positionalities of listeners. This means recognising and supporting all the different ways participants may listen and therefore amplify the participant's role and status in knowledge production. By emphasising *with*, I protect the space for participatory forms of research that are person-centred and value different knowledges. But participatory listening research has the potential to go further. Co-analysis could bring researchers and participants together to interrogate in more detail what is shaping the ways both are listening in the research. Within the parameters of this book chapter, this requires asking you, the reader, to embark on self-reflexive questioning about what is shaping your listening. As I listen, think, and write at my desk, you will engage with these listening prompts and stimulus in a different place and time. But your positioning is more remote than those taking part in this PhD Covid-restricted project.

In the next sections, I invite you to try out the methods described above through choosing either a listening walk or listening-at-home activity. Due to the restrictions of a book chapter, you do not get to share the capture of your listening experience with me. Instead, I offer up exercises that might help you to do your own analysis of how you are listening to your surroundings, what is shaping this listening experience, and the degree to which gentrification may be audible or not. I use this PhD methodology to spark your own sonic understanding of gentrification processes that may be resonating in your surrounding environments.

Listening to Gentrification

(a) Pre-listening exercise questions (make notes if you find this helpful):
   1. What does your street and neighbourhood sound like?
   2. What do you expect to hear in your street and surrounding neighbourhood today?
   3. How do you think it might have sounded in the past, before you lived there?

4. How do you think it might sound in five years' time?
(b) Listening exercise:
   - Choose: either sit near a window and listen for minimum 10 minutes (listening-at-home) OR decide on a short 15–20 minute walk starting at your home (listening walk).
   - Carry out: silently listen at home or whilst walking this short route.
   - Repeat this: now capture the main sounds you hear whilst listening at your window or walking. You can use written notes, drawings, photos, audio, or video recordings. If walking to a destination, you can do this as you return from you walk. Or, if it is a circular walk that takes you back to your home, repeat the walk to do this capturing part.
(c) Post-listening exercise questions:
   1. How do the sounds you captured compare to your answers to the pre-listening questions?
   2. Were there any surprises?
   3. Did you think of any memories or personal stories when you were listening?
   4. How did the listening experience make you feel about your street and neighbourhood?
   5. How did you find this whole exercise?

Listening, both in this chapter and in the research project, has been framed around gentrification. Gentrification is about change. There are myriad debates about how to quantify and measure these changes and prove gentrification is occurring in a specific area (Easton et al., 2020). This includes investigating who is moving in and out, evictions from redevelopments, house price changes, changes in retail and other associated financial flows of money and capital (ibid.; Reades et al., 2022). But what does this mean in listening terms? How can we listen to change? Despite being predominantly considered in visual terms, gentrification transformations have a sonic dimension that some urban scholars have started to interrogate as part of the social and political life of cities (Lisiak et al., 2021; Martin, 2021). As put by one research participant, Desdemona, it is challenging to consider 'what gentrification might sound like.'

Does the above type of listening exercise generate knowledge about gentrifying change? I intentionally left the listening questions open without prescribing the 'sounds of gentrification.' This is not about providing an index of gentrification acoustics to prove whether gentrification is occurring or not. Instead, it is about using listening to stimulate reflections about changes in a surrounding environment. Within a participatory listening research approach, it is about what you think is significant in your street and neighbourhood. The pre-listening questions above were provided to direct your attention on change. These pre- and post- questions are intended as prompts, which are experimental due to the absence of a researcher asking questions during and after the activities. Our remote positionings leave me only with speculations about where you might be listening. You may have been resident where you are listening for a long time and be very aware of how it is changing. You may have identified gentrification processes occurring already. Or you may feel that the opposite is happening, that your surroundings are neglected and declining, causing people to move out. In contrast, you may have not noticed any changes before. You might not have lived there for very long or may not be able to or intend to stay. You may not have known much about the place before you moved there or find it hard to judge how it will change. The interesting part is how you experienced these questions when listening and, through listening, how this might have focused your attention anew onto these issues. In this section, I discuss how the research participants were stimulated to reflect on changes around them through listening in this way. I focus on one cluster of sounds identified by participants that stimulated discussion about gentrification: the sounds of scaffolding and construction. You may or may not have heard these sounds whilst listening, but I use these examples to draw out key points about the potential connections between listening and gentrification.

The three neighbourhoods in my research had already been scoped out as places where gentrification processes were potentially occurring, being prominently discussed in research, media, and policy documents (Prosser, 2022). Brighton, as the benchmark for seaside success, is often proclaimed as a site of mature gentrification, experiencing several waves of gentrification. St Leonards-on-Sea has been identified in a previous study as experiencing early stages of gentrification, which has become a contested issue across Hastings and St Leonards (Shah, 2011; Steele, 2022). In Worthing, there is speculation over the demographic changes and in-migration of families from Brighton and London, impacting its image as a retirement destination, as 'God's waiting room' (Prosser, 2022). These sites differ from the global city neighbourhoods within which gentrification is predominantly researched, including those investigating its sonic dimensions (Martin, 2021; Summers, 2021). Significantly, participants came into the study with different thoughts about whether they felt gentrification was happening around them. For some, witnessing gentrification changes and feeling the effects motivated them to take part in the study. Others were interested in understanding if gentrification was occurring or felt this was misplaced. However, through analysis across all of the participants' material, I found that gentrification was permeating their seaside neighbourhood lives. Participants took different positionings in relation to these changes, but overall gentrification processes were resonating in a number of different ways.

One significant resonance across the three neighbourhoods that amplified the processes of gentrification was the sound of scaffolding and, more broadly, the sounds of construction. Scaffolding was strikingly observed by participants as a 'returning sound' after the strict spring 2020 lockdown. 'Returning sounds' were specific to the particular pandemic conditions of this research project. Residents tried to make sense of 'the pandemic times' through their listening and often commented how sounds that had been absent were now being reintroduced. As stated by Brighton resident Tim:

> What is very noticeable now is all the scaffolding and the builders are back. For a while there was hardly any of that, which is very unusual ... You know, it's a noisy business putting up scaffolding. The whole seafront is all, the buildings are about 200 years old so it's not surprising they need a lot of work.
>
> *(Listening Walk 05)*

Participant descriptions of scaffolding poles clanging, clamps being drilled, and scaffolders calling out to each other is a vivid collection of sounds. As described by Thorin in St Leonards, 'scaffolders, you can always hear their ratchet gun somewhere in the area' (Interview 22). Residents captured and discussed these sounds in different ways. For example, Joan took 14 photos of scaffolding on her walk around Kemptown (Brighton), 'to document every instance of scaffolding' (Interview 08). The sounds of scaffolding returning to the neighbourhood commonly went hand-in-hand with observations about constant neighbourhood renovation. Several commented on how jarring these sounds appeared after a period of relative quietness. Llew, a Brighton resident listening-at-home, found the sound frustrating but also commented, 'I suppose it's a sign that people are back to normal and work is continuing despite the coronavirus' (Listening-at-Home 04). The apparent mundanity and everydayness of scaffolding is significant when thinking about gentrification. Constant renovation and the upgrading of a neighbourhood is a common motif of gentrification. Scaffolding is an indicator of the 'poshing up' of a place, using the aforementioned definition of gentrification given by participant Dr X. Gentrification can be understood through the physical changes that are brought about by reinvestment in the built environment, accompanying a change to higher socio-economic status land-users (Warde, 1991; Clark, 2005). Crucially, as a 'sound stimulus,' scaffolding triggered participants to reflect on why there was so much scaffolding in the area. The contrasting lockdown soundscapes heightened their awareness of the sounds of construction, when previously it might have gone unnoticed. Scaffolding, ongoing investment, and renovation therefore appear to be a 'normal' (Llew, Listening-at-Home 04) part of everyday life in these neighbourhoods. These returning sounds became audible after the strict lockdown had disrupted 'the quotidian soundscapes of capitalism' at the seaside (Yoganathan, 2022: 479).

Did you hear the sounds of scaffolding in your listening activity or other sounds of construction? Have you heard these before in your street or neighbourhood? If so, why do you think you hear these? If not, why do you think these are absent in your area?

Participants had several answers to this question of why the sounds of scaffolding were commonly heard in their neighbourhoods. The reasons given for these constant sounds represent an interesting entanglement of changing urban seaside gentrification features. Many explained that the salty wind conditions require extra up-keep of the Regency architecture; as noted by Jane in Brighton:

> You know they constantly have to be repainted, you know balconies need to be fixed, erm. You know, like a lot of them have those canopies don't they? You know, like those wrought iron railings those fancy canopies and stuff, and they just kind of need to be maintained all the time. But I think the other aspect of the constant building work, the renovations is, erm, with the high turnover you know, Kemptown, I suppose Brighton generally, is that, it's, it's, what do they call it? A transient, people refer to it as a transient city, don't they? (Interview 01)

Jane refers here to the specific combination of maintaining historic seaside resort architecture and a highly transient population. The reputation of Brighton as a transient city is a motif of the seaside attributed mainly to tourism seasons but also part of the decline experienced by many seaside towns. These conditions increase the number of migrant workers living alongside precarious inhabitants navigating poverty, which speaks to the complex racial politics of the seaside, narrated as 'White' in dominant portrayals (Burdsey, 2016). Others also attributed constant renovation to landlords upgrading to exploit a profitable rental market, especially in Brighton and St Leonards. A spike in DIY by homeowners post-lockdown was noted by several Worthing residents, attributed as a knock-on effect of more people working from home. Other factors were also noted by residents. Two Brighton residents, Llew and Tim, both explained that grade-listed regulations necessitate regular redecorating of historic seafront flats (Interviews 04; 05). These regulations are part of heritage policies that value historic buildings as a tourist and place-branded commodity (Steele and Jarratt, 2019). Whilst being entangled in the sounds of scaffolding, tourism was found across the three neighbourhoods to play a role in strengthening gentrification (Prosser, 2022). Llew explained that in previous decades there had been no regulations on these properties, but as they became more valuable and tourism was reinvested in the area, such heritage-justified policies took hold (Interview 04). The sounds of returning scaffolding therefore amplify and entangle different environmental, historical, social, and policy features that are distinct to urban seaside gentrification in this project. Other types of construction sounds in the participants' material indicated a range of small-scale renovation, more typical of pioneer or classic gentrification, alongside larger redevelopment projects, representing new-build gentrification (Davidson and Lees, 2005). The sounds of larger redevelopments were often viewed by participants as more problematic than those of smaller-scale renovations. For example, in Worthing, the construction sounds of a luxury tower block being built on the seafront triggered negative reactions from several participants. Controversies over this redevelopment included the following: the construction impacts, ruining the skyline, being unaffordable to existing residents and putting a strain on local services. Dr X was the most expressive in objecting to 'the monstrosity' new-build (Listening Walk 11). At one point during her listening walk, Dr X listened imaginatively and speculated what the promenade would have sounded like in the Regency period. She immediately contrasted these imagined historic sounds with the construction of the tower block that she felt was 'out of place' at the seaside (Listening Walk 11). She associated 'shiny white towers' instead with London, dramatically describing them as 'an ode to everything that's wrong with our current political system' (Interview 11). The sounds of this kind of redevelopment therefore connected to fears over particular types of gentrification associated with London, such as new-build redevelopments and the demolitions of estates.

I share these examples from my research to demonstrate how specific sounds can help us tune into processes of gentrification through the ways participants listened to them. The sounds of scaffolding opened up the distinct way gentrification processes are manifesting in these south coast neighbourhoods. Participants' capture and discussion of scaffolding made audible the ways tourism, seaside heritage, and gentrification processes are entangled. Listening to our environment can therefore make audible the localised aspects of gentrification and the place-specific ways it is experienced by residents. Where you live and listen may not be impacted in the same way by scaffolding and construction as it is in these urban seaside neighbourhoods. Or you may listen to scaffolding differently through this exercise and reading the accounts of these participants. Construction is a noisy indicator of physical change in a neighbourhood, which, when tuned into, immediately raises questions over what is changing, who is changing it, why, and for whom. Crucially, the different ways participants listened to redevelopment is significant. Many raised these same questions and found the sounds objectionable. They felt certain redevelopments, such as the Worthing luxury tower block, were an intrusion into their neighbourhood and indicative of change they felt would negatively impact on the seaside. These ways of listening objectionably to redevelopment were shaped by ideas of what the seaside should be. There was an interesting range of objections intermixed. This included prioritising Regency architecture, resisting overdevelopment at the seaside, negative class connotations with tower blocks, fears over the impact of richer outsiders moving in, and the injustice of profiteering from housing (Prosser, 2022). In contrast, one participant welcomed this Worthing redevelopment as pragmatic and a defining aspect of urban living. This positive attitude to urban change shaped the way he listening to scaffolding, likening it the sounds of masts and living near a quay or regatta (Interview 03).

These different ways of listening to redevelopment reveal two important aspects of listening to our environments that I wish to conclude this section with. Firstly, these examples show how listening can simulate reflections on change, especially through its distinct temporal dimensions. Many participants traversed time through their listening whether through contrasting lockdown soundscapes with returning sounds or imagining what the seaside would have sounded like in the past. Predominantly participants reproduced 'the original seaside' storyline that tells of a White, working class, fishing village transformed through gentry spa resort development and mass tourism, which has been followed by decline then revival and regeneration (Ward, 2018). Listening therefore tapped into participants' sense of the seaside in the past, present, and future. Many drew on seaside histories to understand how they experience it in the present and to inform any future changes. These temporal dimensions of listening to gentrification informed the pre-listening questions above. Your listening experience is a snapshot, or sound clip, into your neighbourhood. But your understanding of this soundbite is shaped by what has come before and what you think might come next. Secondly, and crucially overall, understanding what shapes participants' ways of listening to redevelopment and other aspects of their environment links into the idea of critical listening positionality (Robinson, 2020). Previous experiences and positionings in the neighbourhood and wider society influence the way participants encounter and respond to their acoustic environment. These examples have started to highlight these different responses and positionings to change. The next section delves further into these ideas of listening positionality by looking at how we might listen reflectively.

I invite you to undertake another set of listening questions. I ask you to go back over what you captured when listening in your home, street, and neighbourhood. These questions are designed to focus attention onto reflective listening. They prompt thinking about your own positioning in your surrounding environment. Following these listening questions, I share further examples from participants' experiences of listening to gentrification. I focus on how participants were listening in public spaces in ways that raised ethical questions and tensions about their own positioning in gentrification processes. This leads us into discussion about how we can listen together to injustices associated with gentrification, culminating in a final listening exercise in the conclusion.

Listening Reflectively to Gentrification and Injustices

(d) Listening reflectively questions
Go back over what you captured from your listening walk or listening-at-home exercise and think about the following questions:
1. Why did you pick out those sounds?
2. Is there anything noticeably missing? Who or what are you not hearing?
3. How do you contribute to your street or neighbourhood soundscape?
4. What do you think shaped your listening?

I found that this project's form of participatory listening research facilitated reflections, some very deep and personal, from participants about gentrification and their own positioning and role in these processes. Participants experienced reflective listening when they became aware of themselves or something significant in their acoustic environment that prompted reflections. This practice of listening reflectively was supported by the elicitation element of the follow-up interview. Using the material captured, and sometimes listening-back together, I asked participants why they focused on specific sounds and what this stimulated. I came to realise that the answers to these questions offered above were also significant. Through interview discussion, we sometimes delved into absent sounds (question 2) and their own sounds (question 3). Discussion of the fourth question was more complexly weaved into participants' personal stories and reflections. This included their previous experiences of gentrification in other neighbourhoods and their reasons for moving to the seaside as well as social identity characteristics and intersections, such as gender, race, and class. Through my creative listening analysis, I grappled with what was shaping participants' listening to gentrification. As a Covid-induced method, there was a degree of experimentation, which led me to understand the importance of these listening questions as part of the project's journey. I therefore offer them here in this chapter to you, as a way of analysing your own listening experience.

These questions are connected to the challenges of Deep Listening (Oliveros, 2005) and critical listening positionality (Robinson, 2020). Robinson (2020: 10) asks us to self-reflexively question 'how race, class, gender, sexuality, ability, and cultural background intersect and influence the way we are able to hear sound, music, and the world around us.' These are therefore challenging questions, especially the fourth question. It is unlikely that you can answer this question immediately or even fully. In my creative listening analysis, I was able to start identifying examples of how social identity and previous experiences were influencing the ways participants were listening. But this was not exhaustive; nor was I attempting to undertake critical listening positionality on participants' behalf. The research findings instead demonstrated the potential of listening-with residents to better understand how gentrification is experienced. Listening-with residents makes the 'seasideness' of these experiences audible, which is distinct from other sites of sonic research into gentrification (Martin, 2021; Summers, 2021). Integral to gentrification experiences is how residents are positioned within it, and therefore, the different impacts and interactions with its injustices.

This final section uses examples to draw out key points about how interrogating listening positionalities is an important part of understanding gentrification resonances. Listening-with participants made audible different value judgements, normative assumptions, and ethical positionings. I identified reflective listening practice as a way into these ethical and social justice dimensions. Some participants were directly affected by gentrification, some felt responsible, and others placed themselves at a distance from these processes. It is not possible within this chapter to delve into detailed discussion of all the injustices and ethical dilemmas identified by participants. I therefore focus on some brief examples of how participants were listening in public spaces, which may or may not chime with your own listening experiences.

The listening walks primarily took place in public spaces, traversing streets, parks, seafront promenades, and beaches with only some residents capturing their private domestic spaces at the end or beginning of a walk. Whilst the listening-at-home activities occurred in private spaces, all the listening activities stimulated abundant reflections about public spaces, including the piers, promenades, cafés, pubs, shops, restaurants, artist studios, galleries, and community centres. You may have encountered different public spaces in your listening exercise. Sounds occurring in public space signal who is using these spaces, how they are being used, who is absent and excluded, how interactions change and are contested. These considerations go straight to the heart of gentrification injustices. For example, several participants identified that gentrifying neighbourhoods creates fewer public and cultural spaces that are available for everyone. This reduces the possibility of encountering others, which is claimed to be an attribute of urban public space (Valentine and Sadgrove, 2012). Public space can be considered critical in the defence 'against forces of commodification, privatisation and state interference' (Soja, 1989: 45). Whilst listening in public spaces, the majority of participants discussed the value they placed on living in a diverse neighbourhood. However, it is important to interrogate what is meant by diversity and how it is constructed. The sounds of human interaction were a significant sound cluster across the material. This included participants noting when they heard different accents or languages, which marked out these sounds by class and/or race. For example, Myrtle, in St Leonards, expressed surprise at still being able to hear working class accents in some of the streets (Listening Walk 02). She had expected to hear louder, 'posh or owning class' voices because she keenly felt the effects of gentrifers moving in (Listening Walk 02). Another St Leonards resident, Virginia, also talked about the 'real mix' of 'different speeches and talks and languages' (Interview 15). But she divided this up between 'migrants' and 'DFLs' (Down from London) or 'OFBs' (Over from Brighton) gentrifiers, which mixes up class and racial considerations. Voices were often marked out as different in terms of race or class by participants when listening in public spaces.

> Did you encounter voices in your listening walk or listening-at-home activity? What stood out to and how did you capture this? Were these familiar voices or were you surprised by these voices? Do you consider your neighbourhood diverse?

Despite many participants asserting how they value diversity, further reflections and discussion often showed exclusions in public spaces. The seaside has long been 'imagined, represented, and consumed as 'White spaces'' (Burdsey, 2016: 112). Participants' listening experiences appear to chime with assuming a White default majority. For example, Mary-Jane initially proclaimed her street was a 'good mix,' especially in terms of age and sexual orientation (Interview 06). But when prompted further, she reflected 'erm, we're not racially diverse. But then Brighton isn't very, is it really?' (Interview 06). As well as a lack of racial diversity, Mary-Jane also described the 'last' working class person leaving the street, making it now a 'classic … what I would call, arty middle class' (Interview 06). Other residents often talked about working class people and cultures as part of the past, for example the 'last,' 'traditional,' or at risk of being lost to the future middle-class trajectory of the neighbourhood.

Thus, there is a tension in the way diversity was listened to and discussed by participants in these seaside neighbourhoods. Many participants described diversity in ways that fit a cosmopolitan approach, asserting positive encounters with difference (Valentine and Sadgrove, 2012). And yet, through their listening, they noted the lack of places to encounter difference, often due to retail gentrification. This valuing of a cosmopolitan-style seaside was mainly discussed by White participants, with the participants identifying as Asian, of African heritage, or in mixed parentage families bringing different perspectives. Myrtle describes the racist abuse that her Nepali partner and mixed parentage son received when they first moved to St Leonards, but how this had dissipated in recent years (Interview 02). Bennie described having people yell 'fake Chinese babbling' at her in the streets of Worthing, though she minimises these incidents by calling it occasional and explaining that she 'had worse' elsewhere (Interview 14). Dr X described the different treatment she received as the 'whitest one' in her mixed parentage African-English family compared to her sister and mother (Interview 11). These different experiences chime with Martin's (2021: 105–108) calls for 'intersectional listening' to gentrification, based on her research into the lives and experiences of Black residents and the deep racialisation of sound, noise, and music. Dr X believed these experiences were not 'just about Whiteness':

> Cause I mean, I, you know, I stack shelves for Wilko and I work with a lot [of] Polish girls, um, and they get a lot of attitude from other members of staff. Because when they're all together, they talk Polish. There's people who've complained about that. It just mind boggles me. They can talk in their own language if they want to (Interview 11).

Dr X describes the policing of workers speaking 'their own language' in the workplace, which connects to the way other participants mark out hearing other languages in the streets. Thus, we see in these short examples how listening in public spaces, and specifically to human interactions, opens up complex processes of exclusions. Lisiak et al. (2021: 262) argue that 'exclusions and inclusions operate through and outside language' with racism and xenophobia permeating both verbal and non-verbal urban sounds. In these seaside neighbourhoods, these racial and classed dynamics are entangled with gentrifying change.

Retail gentrification was frequently described by participants as one of the ways that public spaces were changing in their neighbourhoods. The sounds of cafés were recorded by many participants, which stimulated reflections on this aspect. For example, when Desdemona pondered what the sound of gentrification might sound like, she instantly hit on the sounds of granules being knocked out of coffee machine apparatus (Interview 10). Joan in Brighton picked up several café soundscapes as she audio recorded her walk. She also made notes and drawings, including mapping the shops on one street (see Figure 1.2).

Joan described a relatively high turnover of cafés becoming ever more exclusive as they changed ownership, which chimes with others' observations in the area. Through another Brighton listening walk, Jane was sparked into recalling a lost café that had been run by an older gay couple (Listening Walk 01). She described the impact of losing this affordable amenity:

> What I think is really sad, is that, people like me can afford to go and pay £2.50 for a cup of coffee. But there's a lot of older people on a low income that would go there and they would, they didn't just do fry ups, they did like, you know, meat and two veg for lunch whatever and you know, like,

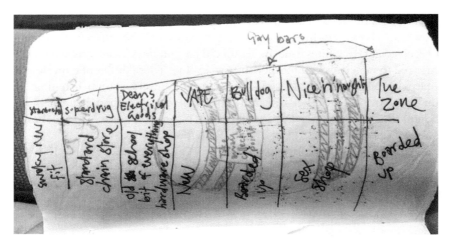

FIGURE 1.2 Drawing by Joan, Listening Walk 08, Brighton.

traditional food. And, you know, if you're an OAP [Old Age Pensioner] and you really don't want to cook and you're on a low income. And you want to get out and be sociable and talk to someone. It's a really important thing (Interview 01).

This story of the lost café shows Jane's sense of injustices playing out in her neighbourhood. But within this concern for others, Jane also distances herself as 'people like me' can afford the more expensive types of cafés. Whilst naming this change and highlighting the injustices, she is not directly affected. Other participants similarly named these exclusionary changes in their listening but reflected on taking part in this shift in café culture. Their navigations of this ethical tension play out in their listening material. For Jane, listening to cafés becomes a story of the lost features of the neighbourhood, which evokes nostalgic sympathy, fixing it in the past and keeping it distant.

These examples from participants listening in public spaces highlight the complex processes of gentrification and associated injustices. Although only touched on briefly for the purposes of this chapter, these examples show the ways public spaces become racialised and classed by participants in their listening experiences. These speak to the work of scholars investigating racialised public space, gentrification, and the policing of neighbourhoods (Gutierrez, 2019; Lisiak, et al., 2021; Martin, 2021; Summers, 2021). Entangled in this gentrifying change, working class people, places, and activities often get positioned in the past as 'traditional,' 'the last,' and 'lost.' This goes back to the listening questions in the previous section that asked what you think your neighbourhood might have sounded like in the past and what it may sound like in the future. For these participants, there are constant comparisons to what their neighbourhood public spaces sounded like in the past and where this might be heading. For you, the reader, you may have a very different understanding of your neighbourhood's acoustic past, present, and future. But crucially this discussion shows how listening to our environments together can prompt us to reflect anew on these dynamics. Some participants were more able to reflect on their own positioning and potential culpability in gentrification, which is touched on in the café example. For example, many participants moved into their neighbourhoods as new owner-occupiers navigating the uneven affordability of the region. But several had also experienced the effects of gentrification from where they were moving, which complicates any ideas of clear gentrification 'monsters' and 'victims' (Prosser, 2022). Listening-with participants to gentrification allowed reflective and in-depth discussion about these nuances and complexities of structural injustices.

It is hoped that these listening exercises open up and stimulate your own reflections on encounters and experiences of gentrification. In the final section, I offer one last listening exercise. This last set of questions asks you to think beyond what you hear in the moment and try to listen imaginatively. This is another experiment in trying to listen deeply and self-reflexively, taking inspiration from both research participants as well as Oliveros (2005) and Robinson (2020). I conclude with reflections on the potential for and limitations to listening to gentrification.

Conclusion

(e) Imaginative listening questions
   Now, imagining beyond your listening experience:
   1. Think of a neighbour or someone you know. Imagine what they would listen to if they did this same listening walk or listening-at-home activity. What do you imagine they would capture? How might this be the same or different from your listening experience? How would it change again if you listened together?
   2. Imagine a street in a society with less injustices, a utopian street. What would it sound like? What would you not hear? What would need to change now to create that soundscape?

This chapter is about being ambitious with the potential of listening to issues of social justice. I started with an experimental invitation to listen together through this chapter. I shared my research methodology and findings on listening-with residents to gentrification at the urban seaside. The above last set of questions offers directions for future experimentation by suggesting ways that we can listen imaginatively with others. In some ways, these are impossible questions – we cannot listen like someone else without our own biases interfering and it is hard to access utopian soundscapes. These are creative prompts, chiming with other contributions in this book, such as the listening pieces offered by Celeste Cantor-Stephens (see Chapter 7).

Overall, I argue that a participatory ethos to listening provides a way forward in these endeavours. This ethos makes a commitment to valuing different ways of knowing and tackling power imbalances whilst being person-centred and inclusive (Beebeejaun et al., 2014; Hall and Tandon, 2017). Through exploring how participatory listening research can be applied to a chapter, two main ways of finding more socially just practices of listening together are made apparent. Firstly, learning from Deaf, queer, and decolonial critiques, it is about embracing and valuing different listening experiences, practices, and positionalities (Haualand, 2008; Bonenfant, 2010; Robinson, 2020; Chaves and Aragão, 2021; Yoganathan, 2022). This includes striving to redress the exclusionary assumptions and tendencies within conventional sound and listening approaches. Secondly, it is about tuning into and actively listening to issues of social justice. This includes using listening to critically reflect and question our own positioning in wider structural injustices, as offered by Robinson's (2020) idea of critical listening positionality. In this chapter, I have focused our attention on gentrification and shared prompts from my findings on how this might resonate around you. Listening-with, as a form of participatory listening research, is offered in this chapter as a way for readers to engage in listening to resonances of gentrification. How did you find these exercises? Did they help you to listen reflectively? What didn't work? How can these be improved or used again, with others, for different topics? Crucially, is listening enough? What are its limitations and what else is needed?

The brief overview of critiques to listening approaches provided in this chapter signal the myriad limitations of listening, which you may have experienced. On its own, listening to our environments, is only one part of the story – a soundbite into the complexities of gentrification processes and how you might sense and experience its associated injustices. As discussed in Robinson's (2020) wider work, a focus on listening risks falling into a traditional Western hierarchy that attempts to separate out the senses. It is perhaps unhelpful to think about ears and hearing. Perhaps instead we should strive to better tune into and understand resonances of social justices. For example, critical disability studies offer work around resonance and multi-sensory practices (McAskill et al., 2021; Sterne, 2022; Drever and Hugill, 2022). What could a participatory sensory approach look, feel, sound, and smell like? What else might we discover by embracing a more encompassing multi-sensory approach? At a minimum, this chapter has aimed to spark curiosity and sonic imagination. The last listening questions challenge us to imaginatively listen to a place with less injustices, to our ideas of utopia. As I write, I find it hard to answer and so might you. But it is hoped that, if we keep experimenting through listening-with others, we might find new reactions, emotions, insights, and eventually actions that bring us closer.

Notes

1 Participants chose their own pseudonyms for anonymity.
2 Deaf is capitalised here to refer to the cultural, linguistic, and historical community of people who have been deaf all their lives, as opposed to describing an audiological condition without the capital D. The critiques mentioned here come from Deaf scholars as well as deaf studies, which primarily focuses on this community of shared identity e.g. Haualand (2008).

References

Alarcón, X. and Herrema, R. (2017) Pauline Oliveros: A shared resonance. *Organised Sound*, 22(1): 7–10.

Arkette, S. (2004) Sounds like city. *Theory, Culture & Society*, 21: 159–168.

Atkinson, R. (2015) Losing one's place: Narratives of neighbourhood change, market injustice and symbolic displacement. *Housing, Theory and Society*, 32: 373–388.

Beebeejaun, Y., Durose, C., Rees, J., Richardson, J. and Richardson, L. (2014) 'Beyond text': Exploring ethos and method in co-producing research with communities. *Community Development Journal*, 49: 37–53.

Bonenfant, Y. (2010) Queer listening to queer vocal timbres. *Performance Research*, 15: 74–80.

Brown, J. and Kirk-Wade, E. (2021) *Coronavirus: A history of 'Lockdown laws' in England*. House of Commons Library, Research Briefing No. 9068, available at: https://commonslibrary.parliament.uk/research-briefings/cbp-9068/. Accessed 28 July 2022.

Bull, M. (2018) Introduction: Sound studies and the art of listening. In: M. Bull. (ed.) *The Routledge Companion to Sound Studies*. London: Routledge, xvii–xxxii.

Burdsey, D. (2016) *Race, Place and the Seaside: Postcards from the Edge*. Basingstoke: Palgrave Macmillan.

Butcher, M. and Dickens, L. (2016) Spatial dislocation and affective displacement: Youth perspectives on gentrification in London. *International Journal of Urban and Regional Research*, 40: 800–816.

Carmona, S.J. (2020) Silences and policies in the shared listening: Ultra-red and Escuchatorio. *Sound Effects*, 9(1).

Chaves, R. and Aragão, T.A. (2021) Localising acoustic ecology: A critique towards a relational collaborative paradigm. *Organised Sound: An International Journal of Music Technology*, 26: 190–200.

Clark, E. (2005) The order and simplicity of gentrification: A political challenge. In R. Atkinson, and G. Bridge. (eds.) *Gentrification in a Global Context: The New Urban Colonialism*. London and New York: Routledge.

Couture, S., Sterne, J., Sawhney, M., Jordan, J. et al. (2020) Sensate sovereignty: A dialogue on Dylan Robinson's Hungry Listening'. The Culture and Technology Discussion Working Group, *Amoderns*.

Davidson, M. (2009) Displacement, space and dwelling: Placing gentrification debate. *Ethics, Place & Environment*, 12: 219–234.

Davidson, M. and Lees, L. (2005) New-build 'gentrification' and London's riverside renaissance. *Environment and Planning A*, 37: 1165–1190.

Dreher, T. (2009). Listening across difference: Media and multiculturalism beyond the politics of voice. *Continuum*, 23(4): 445–458.

Drever, J.L. (2013) Silent soundwalking: An urban pedestrian soundscape methodology. AIA-DAGA 2013, The Joint Conference on Acoustics, European Acoustics Association Euroregio, 39th Annual Congress of the Deutsche Gesellschaft für Akustik and the 40th Annual Congress of the Associazione Italiana di Acustica. *Merano*, Italy, 1–3.

Drever, J.L. (2019) 'Primacy of the ear' – But whose ear?: The case for auraldiversity in sonic arts practice and discourse. *Organised Sound*, 24: 85–95.

Drever, J.L. and Hugill, A. (2022) Aural diversity: General introduction. In J.L. Drever and A. Hugill (eds.) *Aural Diversity*. Abingdon, Oxon; New York: Routledge, pp.1–12.

Easton, S., Lees, L., Hubbard, P. and Tate, N. (2020) Measuring and mapping displacement: The problem of quantification in the battle against gentrification. *Urban Studies*, 57(2): 286–306.

Elliott-Cooper, A., Hubbard, P. and Lees, L. (2019) Moving beyond Marcuse: Gentrification, displacement and the violence of un-homing. *Progress in Human Geography*: 1–18.

Etherington, K. (2004) *Becoming a Reflexive Researcher: Using Our Selves in Research*. London: Jessica Kingsley.

Feld, S. and Brenneis, D. (2004) Doing anthropology in sound. *American Anthropological Association*, 31(4): 461–474.

Friedner, M. and Helmreich, S. (2012) Sound studies meets Deaf studies. *Senses & Society*, 7: 72–86.

Glass, R. (1964) *London: Aspects of Change*. University College London, Centre for Urban Studies, London: MacGibbon & Kee.

Guillebaud, C. (2017) Introduction: Multiple listenings: Anthropology of sound worlds. In C. Guillebaud. (ed.) *Toward an Anthropology of Ambient Sound*. Cornwall, UK: Routledge, pp. 1–18.

Guiu, C. (2017) Listening to the city: The sonorities of urban growth in Barcelona. In C. Guillebaud. (ed.) *Toward an Anthropology of Ambient Sound*. Cornwall, UK: Routledge, pp. 176–191.

Gutierrez, A. (2019) Flâneuse>La caminanta. *Sounding Out*! 12 August, available at: https://soundstudiesblog.com/2019/08/12/flaneusela-caminanta/. Accessed 27 April 2023.

Hackworth, J. (2002) Postrecession gentrification in New York City. *Urban Affairs Review*, 37: 815–843.

Hall, B.L. and Tandon, R. (2017) Decolonization of knowledge, epistemicide, participatory research and higher education. *Research for All*, 1(1): 6–19.

Harold, G. (2013) Reconsidering sound and the city: Asserting the right to the Deaf-Friendly City. *Environment and Planning D: Society & Space*, 31: 846–862.

Haualand, H. (2008) Sound and belonging: What is a community?. *Open Your Eyes: Deaf Studies Talking*: 111–123.

Helmreich, S. (2010) Listening against soundscapes. *Anthropology New*, 51: 10–10.

Herlihy, P.H. and Knapp, G. (2003) Maps of, by, and for the Peoples of Latin America. *Human Organization*, 62: 303–314.

Howes, D. (2005) *Empire of the Senses: The Sensual Culture Reader*. Oxford: Berg.
Lisiak, A., Back, L. & Jackson, E. (2021) Urban multiculture and xenophonophobia in London and Berlin. *European Journal of Cultural Studies*, 24(1): 259–274.
Marcuse, P. (1985) Gentrification, abandonment, and displacement: Connections, causes, and policy responses in New York City. *Washington University Journal of Urban and Contemporary Law*, 28: 195.
Martin, A. (2021) Plainly audible listening intersectionally to the amplified noise act in Washington, D.C. *Journal of Popular Music Studies*, 33(4): 104–125.
McAskill, A., Sawchuk, K. & Thulin, S. (2021) Editorial introduction to VIBE special issue. *Canadian Journal of Disability Studies*, 10(2): 1–6.
Oliveros, P. (1974) *Sonic Meditations*. Baltimore, MD: SmithPublications.
Oliveros, P. (2005) *Deep Listening: A Composer's Sound Practice*. Lincoln, Neb: iUniverse.
Phillips, M., Smith, D., Brooking, H., et al. (2021) Re-placing displacement in gentrification studies: Temporality and multi-dimensionality in rural gentrification displacement. *Geoforum*, 118: 66–82.
Post, J.C. (2006) *Ethnomusicology: A Contemporary Reader*. Abingdon, New York: Routledge.
Prosser, B.M. (2022) *Listening to urban seaside gentrification: living with displacement injustices on the UK south coast*. PhD Thesis, University of Brighton. Available at: https://research.brighton.ac.uk/en/studentTheses/listening-to-urban-seaside-gentrification. Accessed 27 April 2023.
Prosser, B.M. (2023) Researching place during a pandemic: Ex situ listening. *Networking Knowledge: Journal of the MeCCSA Postgraduate Network*, 16(1): 101–117.
Reades, J., Lees, L., Hubbard, P. and Lansely, G. (2022) Quantifying state-led gentrification in London: Using linked consumer and administrative records to trace displacement from council estates. *EPA: Economy and Space*: 1–18.
Robinson, D. (2020) *Hungry Listening Resonant Theory for Indigenous Sound Studies*. Minneapolis: University of Minnesota Press.
Sánchez, I. (2017) Mapping out the sounds of urban transformation: The renewal of Lisbon's Mouraria quarter. In: Guillebaud, C. (ed.) *Toward an Anthropology of Ambient Sound*. NY: Routledge, pp. 153–175.
Schafer, R.M. (1994) *The Soundscape: Our Sonic Environment and the Tuning of the World*. Rochester, Vt: Destiny Books.
Schaeffer, P., North, C. & Dack, J. (2012) *In Search of a Soncrete Music*. Berkeley: University of California Press.
Shah, P. (2011) *Coastal Gentrification: The Coastification of St Leonards-on-Sea*. PhD Thesis, UK: Loughborough University.
Soja, E.W. (1989) *Postmodern Geographies: The Reassertion of Space in Critical Social Theory*. London: Verso.
Steele, J. (2022) *Self-renovating Neighbourhoods as an Alternative to the False Choice of Gentrification or Decline*. PhD Thesis, UK: University of Leicester.
Steele, J. and Jarratt, D. (2019) The seaside resort, nostalgia and restoration – Practising place: creative and critical reflections on place. *Art Editions North*: 132–149.
Sterne, J. (2022) How to hear impairment, at Aural Diversity Network Workshop 4: Soundscape and Sound Studies, 16 Sep 2022, available at: https://auraldiversity.org/workshop4.html. Accessed 27 April 2023.

Summers, B.T. (2021) Reclaiming the chocolate city: Soundscapes of gentrification and resistance in Washington, DC. *Environment and Planning D: Society and Space*, 39(1): 30–46.

Uimonen, H. (2011) Everyday sounds revealed: Acoustic communication and environmental recordings. *Organised Sound*, 16: 256–263.

Valentine, G. and Sadgrove, J. (2012) Lived difference: A narrative account of spatiotemporal processes of social differentiation. *Environment and Planning A*, 44: 2049–2063.

Wang, C.C. (2006) Youth participation in photovoice as a strategy for community change. *Journal of Community Practice*, 14: 147–161.

Ward, J. (2018) Down by the sea: Visual arts, artists and coastal regeneration. *International Journal of Cultural Policy*, 24: 121–138.

Warde, A. (1991) Gentrification as consumption: Issues of class and gender. *Environment & Planning D: Society & Space*, 9: 223–232.

Westerkamp, H. (2017) The practice of listening in unsettled times. In R. Castro. (ed.) *Invisible Places: Sound, Urbanism and Sense of Place*. São Miguel Island, Azores, Portugal, pp. 29–45.

Wrightson, K. (1999) An introduction to acoustic ecology. *Journal of Electroacoustic Music*, 12: 10–13.

Yoganathan, N. (2022) Sounding out normative and colour-blind listening in acoustic ecology. *Canadian Journal of Communication*, 47(3): 462–487.

2

'MADE IN LDN'

Young People's Production of Rap Music in the Neoliberal Youth Club

Baljit Kaur

Shortly after its opening in 2014, I began volunteering at Bass Youth Club,[1] a multi-million-pound creative youth space in East London. As most volunteers generally did, I spent my time there supervising young people in the games room and admittedly being defeated at numerous games of table tennis. A few years later, it was my curiosity to understand young people's experiences of violence and accessible therapeutic interventions that led me from the games room to observing and learning about the production of music in the youth club's music studio. The studio was a space that brought young people together, and where they narrated their lives over what producers referred to as type beats – in other words, they rapped over an instrumental. These observations led to the start of a PhD followed by 10 months of fieldwork at Bass Youth Club between June 2019 and March 2020. It is through this research I sought to explore whether the production of music could be a viable community intervention for young people who have lived experiences of violence.[2]

This chapter derives from the ethnographic research conducted and some of the findings that emerged during this time. As a marginalised population, the voices of working-class young people are often left unheard. Thus, underpinned by a motivation to amplify the voices of my interlocutors and illuminate the various forms of violence they have experienced, I have used multiple quotes throughout taken directly from my interviews. Furthermore, the chapter seeks to highlight the role that Bass Youth Club plays in supporting and shaping rap cultures. It does so particularly against the backdrop of the government's 'Levelling Up' agenda (2002), following a decade of austerity policies. Starting with Grime, a significant Black music genre that has been linked to East London, I provide a brief overview of its

DOI: 10.4324/9781003361046-3

sonic and lyrical qualities. There is a particular emphasis on Grime artists as organic intellectuals (Gramsci, 1971; Charles, 2018), such that the social commentary embedded in their lyricism is situated in the communities that they are part of. Through this lens, the proceeding section demonstrates the way in which Grime artists expose structural forms of violence and class oppression in an authentic way and thus 'keep it real.'

It is the violence of austere policy-making that has underpinned the drastic reduction of youth services and youth clubs; such spaces have historically played an integral role in the artistic development of widely recognised figures within London's Grime scene (Bramwell and Butterworth, 2020; James, 2021). Despite the deficiency in investment, Bass Youth Club has had copious funding, enabling exponential growth; this can be seen in the development of its own record label, *Bass Soundz*, and a partnership with multi-national corporation, Apple. While such partnerships are considered fundamental to improving the offer for young people, this chapter builds on previous work (Bramwell and Butterworth, 2020) to explore the way in which rap cultures are shaped by public funding. In so doing, it foregrounds the tensions that emerge between 'keeping it real' and 'making it,' two distinct discourses with which rap artists habitually engage, against the backdrop of an increasingly fraying meritocratic society (Littler, 2018; Social Mobility Commission, 2020). It asks what is at stake in the acquisition of commercial success.

Keeping it Real

East London has significant links to a number of Black music genres. In the early 2000s, it is where Grime's first wave of MCs (rappers) and its pioneers started out (White, 2017; Charles, 2018; James, 2021). My interlocutors Rick and Damien, two Grime artists and music producers who were part of Grime crews, identified the artist, Wiley, as the genre's innovator, emerging in the East London boroughs of Bow, Tower Hamlets, and Newham. Written rhymes and spitting (rapping) over Jungle evolved into a fiercer and stripped-down sound often made on computers and 'cracked' software to produce Grime (Adams, 2019). Production software programmes like Logic and Fruity Loops, which came with a standard 140bpm tempo shaped its sonic qualities making it notably faster than UK Garage and slower than Jungle (Adams, 2019; James, 2021). As Damien explained, 'it wasn't called Grime at the time, but it was much rougher, much harder … had much more pace to it – had much more energy.' Moreover, Grime's lyricism is significant for the conditions through which it emerged. It is a working-class scene, originating in the council estates of East London where 42% of London's social housing was and where many Grime artists grew up (Perera, 2018). It also coincided with New Labour's plans to regenerate the 'impoverished edge of a wealthy city' (Hancox, 2018: 18). In addition to the development of new luxury apartments, urban regeneration projects were accompanied with heightened CCTV surveillance and intensified punitive measures in the form of anti-social behaviour orders.[3] These served to criminalise non-criminal behaviours in Black working-class neighbourhoods (Perera, 2018). While Canary Wharf, at the time, became the largest urban regeneration project in the world, the 'second city,' as it later became known, was never designed to have a relationship with the council estates situated less than two miles away. Articulated by Grime artists, Tinchy Stryder and Dizzee Rascal, who as children lived on the Crossways Estate: 'everything felt fresher and cleaner than where we grew up' (Tinchy Stryder, 2018: 23), 'there are rich people moving in now […] you can tell they're not living the same way as us' (Dizzee Rascal, 2018: 23). Through this lens, both Tinchy Stryder and Dizzee Rascals' comments are indicative of the structural problems of inequality, poverty, racism, and exclusion embedded in the fabric of East London.

In this way, Grime wasn't about 'showing off, nor flossing,' rather, getting 'straight to the point, no fucking about' (James, 2021: 93). Championed as the 'Godfather of Grime,' Wiley states that the 'sound came from our situation. It's a cold, dark sound because we came from a cold dark place. These are inner-city London streets' (Wiley, 2017: 79). It was the sound of the lived experiences of 'the best MCs' as Wiley asserts, 'from when they were down and never had a fiver' (Wiley, 2017: 297). This authentic realism that reflects the significant realities of social inequality could be heard in earlier productions of Grime tracks. In the 2004 track, 'Graftin', Dizzee Rascal, one of the pioneers of Grime, makes references to teacups, red telephone boxes and Buckingham Palace as quintessential markers of London, and contrasts this with the visual backdrop of a council block while rapping about the grittiness of East London (Charlie Kinross Producer/Director 2007). The juxtaposition of Buckingham Palace and the council block serves to convey the conditions of urban dwelling, that ultimately alludes to experiences of structural violence. Using Galtung's (1969) definition, it is this violence that is routine and silent and as 'natural as the air around us' (Galtung 1969: 173). It is 'the violence of injustice and inequality' (Rylko-Bauer and Farmer 2016: 47) through which harm is created by 'preventing people meeting their basic needs' (Cooper and Whyte 2022: 2010).

Grime is thus considered a politically significant subculture through which working-class young people document Black diasporic life (James, 2021), lending to the autobiographical nature of rap music. In US literature, the notion of 'keeping it real,' or 'the real deal' as Perry (2004: 90) argues, pertains to an expression of the problems of the 'ghetto' – telling one's story of loss, sorrow, exploitation, rage, and despair about conditions that are both denied and neglected in wider society (Rose, 2008). It is music that provides an alternative public realm for Black diasporic expression in the face of systemic discrimination, resonating with issues of authenticity and resistance

(Basu, 1998). A way of being true to yourself, 'a voice by the people for the people' (Basu, 1998: 375). In the context of East London, where regeneration is understood as a response to grit and disorder, Grime is considered an oppositional force, a cultural expression through which Grime artists present a localised interpretation of the Black diasporic experience: 'what you see when you wake up in the morning … a lot of grime in the area, a lot of grimy things happening' (Hancox, 2018: 18).

In light of these narrations, Grime artists should be understood as organic intellectuals (Charles, 2018). Although such artists are often found at the margins of the music industry, they are 'rooted and routed' (Perera, 2018: 88) in their communities, therefore holding powerful positions as social commentators. Furthermore, while their lyrical content is not always explicitly political, the lived experiences of poverty and adversity are, and are imbued within the lyrics. Thus, in holding this position, Grime artists have become the voice of their communities, connecting individual frustrations to a collective experience (Charles, 2018). The political power of Grime artists can be seen in the example of Generation Grime. Charles (2018) draws on the 2017 election campaign during which Grime artists used their positions as organic intellectuals to galvanise Generation Grime into political action, largely through social media. The use of technology – hashtags and viral social media content, specifically – enabled social media clicks and shares to be transformed into political action. Young people voted Labour in significant numbers, which had an impact on the outcome of the election (Charles, 2018). Crucially, young people voted in response to their lived experiences, the government's disregard for their lives and for their future, and the disconnection they felt with the Conservative party as knowing little about the localities that young people call home. In this way, such tactics were a means of reaching out to young people and engaging with formal politics. It was a popular movement pushing for social change *with* the collective, and *for* the collective (Charles, 2018), echoing Basu's (1998) articulations of 'keeping it real.' The potential for political engagement and political action in the pursuit of social change is thus exemplified through Generation Grime. However, to consider the role that youth clubs play in the shaping of rap culture, the following section proceeds with an overview of youth clubs, and the impact that funding, and lack thereof, has had on youth programmes.

Youth Clubs and the Violence of Austerity

Historically, youth centres have played an integral role in the artistic development of young people by providing free access to space and technology. Rapping into a microphone, the recording and production of a track makes technology a central component to contemporary Hip Hop and Grime culture. In such institutions, widely recognised figures within London's Grime scene such as Tinchy Stryder, Dizzee Rascal and members of Ruff Sqwad attended Linc Centre in Tower Hamlets (James, 2021), and could be found honing and polishing their skills and practicing the 'tightness' and delivery of their lyrics. In this way, the provision of space, resources and staff have contributed to nurturing a vibrant rap culture in East London (Bramwell and Butterworth, 2020).

Yet, throughout its history, youth clubs have faced numerous challenges largely revolving around the reduction in public expenditure (Ord and Davies, 2022). Many of these challenges were introduced under Margaret Thatcher's (1979–1990) government, wherein neoliberal characteristics of privatisation and deregulation shaped the way in which youth services would operate (Bunyan and Ord, 2012; Ord and Davies, 2022). Under Tony Blair's (1997–2008) New Labour government, responsibilities for children and young people's services, how they would be funded and delivered, and by whom, continued shifting, prioritising partnerships and collaboration (Davies, 2019). A corollary of this was competitive procedures of commissioning, procurement, outsourcing, and contracting (Davies, 2019; White, 2020). Over the last decade or so, this has been the principal idea underpinning austere policy-making. Between 2010 and 2016 – the period of David Cameron's Conservative leadership – nearly every local authority faced the brunt of the government's austerity agenda, totalling a sum of approximately £387m in cuts from youth service spending across the UK (Unison, 2016). These cuts particularly impacted young people from poorer backgrounds, young Black people, young LGBT people, and young women; those who needed youth services the most (Unison, 2016). The detrimental consequences felt by local authorities epitomises the Cameronism ideology of 'do[ing] more with less' (O'Hara, 2014: 10), leaving severely neglected youth services to further deteriorate (Ord and Davies, 2022).

Moreover, Green London Assembly Member, Siân Berry, reveals the way in which austerity policies have intensified the detrimental impact on council budgets in recent years. Key findings from her report (2021) indicate that over £36m was cut from annual youth service budgets between 2011–2012 and 2021–2022 across London. This totals a deficiency in investment in young Londoners of over £240m. Consequently, from an initial 300 youth centres across London, 130 have been forced to close, hand in hand with a reduction of over 600 full-time youth worker jobs (Berry, 2021). Through a closer lens, of the further £3.4m budget cuts to be made across London in 2021–2022, the most substantial has been made in Sutton, South London, which closed its last remaining facility in 2021 (Berry, 2021). Sutton joins Waltham Forest, East London, which has been without open-access youth centres[4] for years. Despite no further cuts to be made in Newham, this borough is said to have been the worst affected (Berry, 2021). In 2015–2016, Newham's Shipman youth centre for example, was only one of Newham's four youth centres to have a recording studio in use. However, the loss of skilled youth workers has forced its closure (Bramwell and Butterworth, 2020). Given that numbers have plunged steeply, and the current plight of youth clubs, Ord and Davies (2022) write that it would not be surprising if many young people in the current teenage generation no longer knew what a local youth club is, or that such a facility once existed.

Alongside the reduction of youth clubs, it is also crucial to note the way in which youth work practices and programmes have subsequently changed. The rollback of the welfare state since the 1970s was met with the challenge of bringing a new 'workless class back into society' (Tyler, 2013: 159; Blair, 1997) via employment. Economic inequalities were considered a result of poverty of aspiration and an unwillingness to grasp opportunities (Tyler, 2013). During the 1990s, the fiscal imperatives of neoliberalism were made further visible (Tyler, 2013); laws and regulations monitored and restricted young people's use of public spaces (Turner, 2017). Two decades later, David Cameron spoke from a community centre in Barlanark, the deprived East End of Glasgow. There, he declared he would 'repair our broken society' (Tyler, 2013: 176), from crime, social disorder, and deprivation. Social problems compounded by austerity measures, such as the increase in child poverty, material deprivation, youth unemployment, and the damage caused to young people's life chances were deemed 'consequences of the choices that people make' (Tyler, 2013: 176; Cameron, 2008). This not only echoed similar sentiments of personal responsibility as those of his predecessors but continued to obscure the structural violence that pervades the lives of working-class young people.

In turn, guided by a neoliberal policy framework, youth work practice became oriented towards crime prevention and concerns around anti-social behaviour, and the programmes offered were designed around behaviour modification (Turner, 2017; Bramwell and Butterworth, 2020). Indeed, this also applies to music programmes whereby rap is used as an 'engagement tool' to cultivate 'ethical citizens' (Bramwell and Butterworth, 2020: 179). Such programmes are underpinned by neoliberal logics of developing rational autonomous subjects, interwoven with the process of self-making through which a path out of poverty can be attained (Tyler, 2013; Ganti, 2014). Transformed by this neoliberal turn, the role of youth work has changed to promote self-reliance and enable responsibility among individuals under the guise of projects of the self, rather than projects of the social (Garasia et al., 2015). As an inherently Conservative pursuit, youth work practice has been argued to centre a focus on the 'moral ills of individual young people, rather than generating social change' (Garasia et al., 2015: 3), so that they can fit into the established social order. In the pursuit of finding biographical solutions to systemic contradictions, youth work practice has the potential to deny, if not overlook, the impact of social problems on young people (Garasia et al., 2015). It is through this lens that Bass Youth Club's role in supporting and shaping rap culture is discussed, with consideration to its very own record label, Bass Soundz, and the youth club's partnership with Apple.

Bass Youth Club

The borough in East London where Bass Youth Club is situated has been subject to substantial cuts in the last decade. Despite such challenges, at the time of the fieldwork, I was told by one of the Service Managers, Alex, that the youth club had a 'healthy bank balance.' A £7.5m budget granted the construction and development of its world-class creative youth facility and world-class programmes.[5] Its three main floors centred around a 260m² triple height performance hall, constituting what its architects have described as a piece of theatre. The youth club brings young people together to get creative, get active, and get inspired; it is a 'place of specialism' Alex told me, where the creative arts are celebrated and made accessible.

Bass Youth Club's exponential growth over the years from one centre to six satellite centres meant that they were now running a service rather than a centre. At the outset, to create safe spaces for young people to chill out, the local council provided funding for three youth workers to open the youth club for three hours, five days a week. It was also funded by an award-winning housing and regeneration community association in East London. At the time of fieldwork, this Housing Association owned 9,000 homes across two wards in the borough, comprised of social housing, shared ownership, and private landlords. As a result, the income generated was reinvested into services to support the community, such as the youth club.

The multi-million-pound investment could be seen in the youth club's facilities comprised of a theatre hall, games rooms, a radio room, gyms, a boxing ring, and a media suite. Situated on the first floor was a music studio where young people participated in music programmes that ran in the evenings, five days a week. The music studio consisted of six small recording rooms and a central space where young people gathered at the start of a music session. In this central space were two keyboards, speakers, a drum kit, and two Apple Mac computers, which were used by music producers and facilitators to demonstrate tasks and record tracks. Each recording room had a microphone, headphones, and an Apple Mac computer from which young people could record tracks and access their projects. During the sessions, the recording rooms were engulfed with the overlapping sounds of type beats downloaded from YouTube, rehearsals of bars (lyrics) that were repeated to perfection, laughter, and banter. Each room was separated with a soundproof glass window. In the central space, attendees were routinely encouraged to perform their tracks to a crowd of young people and often youth workers; it was an opportunity to develop their confidence and their performance skills as aspiring music artists.

Bass Soundz and Apple

Emerging from the continual quest to improve the offer for young people was Bass Youth Club's very own record label, Bass Soundz. The label emerged from humble beginnings after Oscar, Lawrence, and Chris who had been facilitating music programmes since the youth club's opening, noticed that there were 'young people who wanted a bit more time, a bit more energy, advice and guidance as to how to shape their music and make it into a feasible career.' As well as mirroring other record labels in terms of governance and monetary reward, Bass Soundz was also driven by competition. In conversation with Oscar, I was told that it was an opportunity for young people to be given guidance as to how they could 'improve their musicianship, their song-writing, their collaborative skills – how they can work together.' The label took on ten artists or acts per season and each season ran for one year. The limited number of spaces lent to a rigorous selection process. Young people were observed and considered for the record label based on the prerequisites of stage performance, performance capabilities, confidence, lyrical content, marketability, ability to generate followers, promote their work and 'create a bit of hype around what they do.'

To prove themselves as viable commercial music artists, young people were asked to attend a variety of music programmes during which facilitators could 'sit down with them and discuss their ambitions and their musical plans.' This 'progress plan,' as Oscar described, efficiently recorded their musical development, and if they were good enough, they were given the 'final tick.' While the selection process was 'difficult because you want to help everybody, and you want to be able to work with artists who are very strong and very head strong, and have a really strong work ethic,' Oscar asserted, 'we all wanna listen to music that's good, so I guess the most fundamental thing is their output quality.' Thus, in order for young people to be 'fit for the label,' they were encouraged to perform at events such as established festivals, or at venues like the Apple store in Covent Garden, through which they could embody the 'higher tier' artist and demonstrate the 'talent to back it up.' Bass Soundz was therefore one way in which young people were being taught about commercial success. The dissemination of their music through a record label as opposed to local ventures was indicative of the commodification of these young people as music artists. Additionally, this was interwoven with developing rational autonomous subjects, such that young people were selected on the basis of a good work ethic, proactivity, and productivity. In cultivating a sense of personal responsibility, young people could demonstrate that they were 'working hard and getting on' (Dabrowski, 2021: 92), thus conforming to the neoliberal matrix.

Despite their healthy bank balance, as with other youth clubs in East London, Service Managers often respond to the dramatic cuts to local government budgets by developing partnerships. This enables other people to come in and deliver a service to young people, which ultimately expands and enriches the offer. In light of Bass Youth Club, improving the offer meant funding specific programmes and team members. As Alex explained, there was a whole other programme that needed to be paid for, and there are 'lots of different funders that we go out to improve the offering, and not just open up a youth club for three hours.' Part of this additional funding was received from the Big Lottery Fund, while others came from the Home Office and the Greater London Authority (GLA). One of the benefits of being a recipient of the three-year GLA funding was the connections and partnerships to emerge from it. Most significant was the youth club's partnership with multi-national corporation, Apple – one of the largest technology companies in the world, alongside Microsoft, Google, Amazon, and Facebook (The Verge, 2022). GLA's partnership with Apple was subsequently brought to Bass Youth Club through the programme's lead contact. For Operations Manager, Stacey, connections with Apple were welcomed because ultimately, 'it's a brand that young people connect with, and um, are interested in.' Compounding the appeal and interest for young people was access to Apple's design and production software; a 'big brand with very expensive products' were considered to be 'inextricably linked with the music industry.' Stacey continued to tell me that 'if you wanna make it in the music industry – I mean, you don't have to have Apple products, but the reality of the music industry is Macs are part of the industry, so I think giving them access to those *is* important.' In a similar fashion, music producer, Oscar, agreed:

> I'm thinking of it purely from [the perspective of] a music head, that [Apple] made the best hardware to be able to record music and it's still a huge part of my life – I've always had that brand and I'm a fanboy [...] the bottom line is, without this technology we wouldn't be able to create music, so yeah, I think young people should be able to go to Covent Garden and go and perform in the Apple store, and understand that even though it's a pricey and quite an exclusive brand, at some point they'll probably have to stump up the money because in the end they will get a lot of value out of that product – out of that equipment, you know [...] I saved up all my Saturday job money to buy an Apple Mac computer so that I could record music when I was 19 years old, and to be honest with you, it's enabled me to create hundreds and thousands of pounds of music over the years, so I wouldn't – I don't regret it, you know.

The importance of access to Apple products as Stacey and Oscar conveyed was reflected in the music studio and media suite, which largely comprised of Mac computers. Installed on the computers was the Apple software, Logic Pro, with which young people could record and produce their music. The retail price of Logic Pro was marketed at just under £200, however, this was made freely available at Bass Youth Club. Access to Apple products was particularly significant against the backdrop of austerity policies that have drastically reduced youth services, access to space, equipment, staff support, and technical knowledge across the last decade. Thus, a partnership with Apple, Stacey asserted, was 'really positive because it can give a lot more access to those products which are essential to do the creative work.'

In more recent ethnographic research, however, partnerships with Apple can be seen to extend beyond enabling creative work. A young male Driller[6] by the name of Chief Keef is one example of a young rap artist who at the age of 16 became an internet phenomenon after uploading a Drill music video on YouTube (Stuart, 2020). Although he lived in one of the most maligned neighbourhoods in Chicago, the success of his music led to features on the forthcoming albums of mainstream music icons, including multiple Grammy Award-winner Kanye West. It also led to an agreement with long-time music producer and music mogul, Dr. Dre, to establish a line of luxury audio headphones called 'Beats by Keef' – a brand currently owned by Apple. In this way, a combination of hard work, talent, and a partnership with a multi-national corporation such as Apple embodied 'endless opportunity in a free market' (Fernandez and Hendrikse, 2015: 1), the ability to achieve commercial success and as such, the American Dream.

Making it – Made in LDN

The 'British Dream' (BBC, 2017: 1), the British equivalent of the American Dream is understood through the language of meritocracy, whereby aspiration, upward mobility and opportunity are available for all to rise through the social structure (Heather, 2012: 1; Littler, 2018). As Apple products have become more ubiquitous, they have been strategically marketed at its future customers: students, artists, and entrepreneurs. Apple also offer youth programmes and, much like Bass Soundz, developing talent is a significant feature. One example of such an Apple programme brought to the youth club was 'Made in LDN,' which launched in Covent Garden in 2019. 'Made in LDN' was a collaboration between the Mayor of London, Sadiq Khan (2016) and 'Today at Apple,' a programme which sought to develop young people's creative talents. The programme's ethos, parallel to that of Bass Soundz, would offer 'young people a crucial stepping stone as they developed their creative careers' (Made in LDN, 2022: 1). Thus, ahead of the 'Made in LDN' launch, Bass Youth Club were approached by the programme's lead contact in search for some young performers to perform at it. Stacey's response was an immediate 'YES!', because 'I was really keen on getting exciting and the best performance opportunities for our young people.' In this way, the opportunity to perform in a 'professional' context in Covent Garden was not only considered a 'big deal,' as Stacey asserted, but a way in which young people could also 'realise their possibilities.'

Part of the endorsement of such programmes and subsequent performance opportunities was to demonstrate a commitment to building young people's confidence, hard work, productivity, and aspirations, as well as giving talented artists exposure to industry professionals and commercial success. In turn, young people could demonstrate their performance capabilities and begin forging a feasible career from their music. A partnership with Apple therefore epitomised the notion of 'making it.' Put another way, it placed greater emphasis on the acquisition of commercial or industry success (Sköld and Rehn, 2007). While young people wanted to feel seen and heard, and 'just feel like someone,' it was also about 'making money any way possible' (Sköld and Rehn, 2007: 56). Young people frequently expressed their desires for global and worldwide recognition and economic success. 'The world will know us,' I was told, 'everyone will know us.'

Desperate to 'make it,' to 'prove everyone wrong' and not accept being 'ordinary' was 18-year-old Leo. In pursuit of his aspirations, a year prior, Leo had taken the bank card of his uncle with whom he lived at the time and booked a ticket to New York. 'I booked a ticket,' he said, 'like, fifteen hundred pounds [...] I had a couple of songs written down and I wanted to make it, so I wrote it down and I took it.' He described being homeless whilst there and told me:

> In my head, I was just fixated on New York. New York's – America's where you make it [...] I went to all these big record labels. I went to Def Jam[7] and all these other places, I just put them down in my phone and I just went to them.

Upon reflection, Leo explained, 'obviously you can't just go to them, they're in these massive buildings and they're not even – they're not just a building, they're on a [particular] floor in a building.' To his dismay, the trip to New York did not follow through as he had envisioned because 'people who come in Def Jam have, like, passes and stuff,' without which, Leo was told, 'go, you can't be in here.' Back at Bass Youth Club, however, Leo had performed at several events including a local community venue, a venue in

Canary Wharf and at an Apple store. The event at the Apple store, he told me, was:

> Mental [...] We went in Adison Lee cabs from [Bass Youth Club] – that day was crazy. We got there really early and then everyone was dressed up. I wore a jacket, like, double denim, denim jacket, denim jeans ... I had a customised jacket because one of my sister's friends, she knew him at Uni, he was an artist and he decorated it for me.

The jacket was embellished with the title of the track he subsequently performed. He continued:

> Apple took us downstairs to some kind of behind the scenes room [...] offered you a drink, packs of water for us and stuff, asking how we are and behind us there was, like, a screen of our names and our Instagram and social medias [...] all the artists were, like, kind of at the front of the audience and we got called up when we were gonna go up next.

An hour into our interview, Leo proceeded to share with me a song that he'd written over the previous three weeks. Whilst it was not the same song that he'd performed at the Apple event, it was nonetheless a reflection of the kind of rap music he had been writing. The track depicted a scene at a club filled with 'Beckys and Keishas.' He describes an interaction with a 'keen' love interest, whose attractiveness is accompanied with the descriptors 'salty' and 'seasoned.' Interwoven with these exchanges between Leo and the love interest are references to his 'jeweller,' who he asks to 'ice my wrist.' 'My diamonds must hit,' he tells his audience, followed by the ad-lib 'that's a must.' In this track, Leo embodied 'that guy, iced up like a freezer.'

As the track finished playing through his phone, Leo told me that the lyrics were based on his experiences of partying and dancing with girls, who he subsequently described as '*proper* keen and eager.' He told me, 'I talk about girls a lot in my track, even though sometimes I don't even know – when I look at my notes, I just see a lot of, like, tracks about girls – and I've only been with *two* girls.' He justified his lyrics by adding:

> A lot of girls they don't really like – they're all into rap stuff, hood stuff, like, gangster raps, that's what they're into. So, you're like, oh, I have to be like that. That's what's popping, that's what's going on, that's what's clicking right now. So, if I'm gonna make it in music, all my friends listen to this music – well, not all my friends but everyone's doing that sort of music, sometimes you do feel that pressure.

In these comments, Leo exemplifies the pressure of producing music that an audience largely comprised of his male friends and girls are going to be 'into.' Indicative of the kind of performative space that rap music provides, Leo can be seen to be 'flexing,' a term used to describe the reinforcement of a certain persona that in this context maintains a hustler mentality. Stuart (2020) documents the way in which ordinary people can create a public persona, produce content, and build popularity among followers and fans; this is particularly heightened through online platforms. In his track, Leo's flexing can be seen in the embodiment of the materialist bravado – a key convention in rap since its inception. The materialist bravado enables artists to claim that they are the best at something; supposed wealth, sexual abilities, and acts of dominance underpinned by violence are particularly prevalent (Oware, 2018). The materialist bravado exhibits hallmarks of what has been described as 'playa rap.' Its lyrics centre 'spectacular consumption' and depicts a 'flashy, over the top style' (Randolph, 2006: 206). Contrary to presentations of violent masculinity, themes of materialism enable presentations of an alternative masculinity of luxury, which has historically been commercially successful in rap.

Leo continues to explain that '[girls] want this sort of *Love Island*[8] type of guy' who seemingly have an affinity towards brands; Gucci belts and Air Force trainers were listed as examples of such desired items. In his references to 'my jeweller,' 'ice' and 'diamonds,' Leo exemplifies the performance of an alternative persona that he finds himself 'feeding into.' Through his lyrics, Leo's presentation of a sexual, and to a greater extent, materialist bravado, demonstrates two things: firstly, through his copious references to girls, he conforms to the internal expectations of hegemonic masculinity amongst his male friends. In this way, conforming and performing for the membership of a peer group allows Leo to 'keep up with the pack' (Alexander, 1996: 142). Secondly, through the presentation of a materialist bravado, Leo cites the commercially successful artists that have come before, and in so doing, serves to generate followers, promote his work, and create a bit of hype, synonymous with the higher tier artists at Bass Youth Club. Responding to the pressures and personal responsibility for young people to acquire economic success, it is ultimately through this performance that Leo is able to produce music that is 'popping' and 'clicking' in the quest to 'make it.'

Aspirations, Opportunities and Levelling Up

Leo's story is significant in demonstrating the tensions between 'keeping it real' and 'making it' at Bass Youth Club's music studio. As the discussion above illustrates, the various narrations of his intimate relationships and embodiment of a materialist bravado through which he tells his jeweller to 'ice my wrist,' supersedes the hard-to-hear truths of his lived experiences.

The preceding narrative is particularly poignant when read against Leo's experiences of financial adversity – the very circumstances that led him to New York, the homelessness he experienced whilst there, and the subsequent move to a hostel upon his return because his uncle was no longer willing to accommodate him. Much like the flexing Drillers that constitute Stuart's (2020) ethnographic study, writing his lived experiences into his music was deemed 'quite embarrassing.' Leo asserted:

> None of my friends live in hostels and not many people I know live in a hostel, and it's just … it's not something I'd wanna put in my song.

Moreover, my dialogue with Leo does not seek to perpetuate the dichotomy between 'good rap' and 'bad rap' (Perry, 2004). Rather, it serves to illuminate the existing tensions between such narratives. It shines light on the choices that young people make about which stories to tell according to how successful or popular they might be, and especially so if they are striving towards the higher tier of artists with opportunities to perform at external venues. This is not to say that stories of their lived experiences were absent in the studio, they did exist. It mattered, however, *who* these stories were told to. They were told to me, for example, during our interviews, but in the process of selecting songs to perform at venues such as Apple stores, the telling of one's story that encompasses narrations of loss and sorrow (Rose, 2008), as it did for Leo, were cast in a less favourable light, one in which artists would be labelled as 'soft,' 'lame,' or considered to be 'talking all mushy.' 'They're gonna laugh at me,' Leo told me, 'they're all gonna laugh at that […] they'll just laugh at it.'

It is also important for Leo's story to be read against the backdrop of perpetual budget cuts. As stipulated in the 'Levelling Up' agenda under Boris Johnson's (2019–2022) Conservative leadership, these cuts are sought to be repaired with the promise of £560m to be invested in improving youth facilities, services, and experiences for young people in England in the spirit of giving 'everyone the opportunity to flourish' (HM Government, 2022: 1). Secretary Michael Gove further adds that 'Levelling Up and this White Paper is about ending this historical injustice […] to break the link between geography and destiny so that no matter where you live, you have access to the same opportunities' (Gov UK, 2022: 1). That said, far from being new money, £500m, or nearly 90% was in effect the third launch of the September 2019 Youth Investment Fund and has been largely unspent (Merrick, 2021; Simpson, 2021; Ord and Davies, 2022). In turn, the delivery of music programmes at Bass Youth Club continue to be underpinned by competitive processes of funding, developed and augmented on the foundation of partnerships. Parallel to the 'Levelling Up' agenda, the language of upward social mobility, aspiration, and opportunity is interwoven with multi-national corporations like Apple and echoed in my conversations with the youth club's management team. For example, Stacey asserted:

> We need to increase young people's aspirations by showing them that these opportunities are there if they go out and get them […] we wanna aim to empower young people – to make them think bigger than what might be outside their immediate mindset. Therefore, if we said 'no' to opportunities to work with brands like Apple, I feel like in my gut, that is wrong.

To reiterate Stacey's point, such opportunities fuelled by Bass Soundz and Apple were presented with the intentions of increasing young people's aspirations and making them think bigger against the backdrop of deepening inequalities. London, for example, has had the highest rate of child poverty than any other UK region for the last two decades (Joseph Rowntree Foundation 2023), and the borough of Rowe has one of the highest percentages of child poverty across all London boroughs (ibid.). In line with neoliberal logics, this preserves the notion of the responsible individual and the effort to 'better oneself' (Spohrer, 2018: 1) whilst structural violence continues to pervade their lives. Contrary to encouraging young people like Leo to expose their social conditions, such music programmes encourage them to 'make it' in a society where money is 'not easy to get' and upward social mobility 'is not a lived reality in the United Kingdom today' (Social Mobility Commission, 2020: 17). Through this lens, a partnership with Apple was 'sort of like dangling a carrot which is unattainable,' Stacey asserted, and rather than fostering a movement of social critique, the neoliberal youth club can be seen to be complicit in the celebration of the status quo.

Conclusion

This chapter draws on the role that youth clubs play in the supporting and shaping of rap culture. While substantial cuts to youth services have cumulatively forced the closure of many youth clubs across London, Bass Youth Club's recipiency of funding has enabled its exponential growth, opening beyond its initial contract of three hours a day, five days a week. Positioning itself as specialist, Bass Youth Club is a place where young people can get creative, active, and inspired, owing to the provision of space, resources, and staff expertise. This also lends to the development of the youth club's very own record label, Bass Soundz, which like other record labels is both underpinned by and drives competition. With just ten spaces per year, young people compete for the opportunity to improve their musicianship under the guidance of the music facilitators. Amongst a range of prerequisites, higher tier artists are fundamentally expected to generate followers, promote their work, and create a bit of hype around what they do.

In addition to the Bass Soundz record label, the youth club's partnership with multinational corporation, Apple, is also a result of additional funding from the GLA. From the perspective of service managers these partnerships serve to expand and enrich the offer for young people. Through Apple's 'Made in LDN' programme, young people are encouraged to perform at events at Apple stores in places like Covent Garden. However, it is also through such avenues that young people not only prove themselves as viable commercial artists, but in line with neoliberal logic, as rational autonomous individuals who are working hard to get on. This is specifically demonstrated through the example of 18-year-old Leo who is desperate to 'make it,' to prove everyone wrong, and not accept being ordinary. Leo's lyricism conveys a sexual and materialist bravado through which he can be seen to flex. It is in these presentations and performances that the tensions between 'keeping it real' and 'making it' are revealed. Leo's narrations of financial adversity and homelessness are superseded by lyrical references to girls and supposed wealth, strategically citing the commercially successful artists that have come before.

In the pursuit of generating followers, popularity, and mass appeal, I return to the question asked at the beginning of this chapter: what is at stake in the acquisition of commercial success? Mirroring the language of the government's 'Levelling Up' agenda, Bass Youth Club's partnerships with Apple may serve to increase young people's aspirations, making them think 'bigger' and providing them with an opportunity to flourish. In so doing, however, young people can be seen to respond to the pressures of 'making it' in an increasingly fraying meritocratic society. In preserving the personal responsibility for young people to better themselves, opportunities for social critique and justice – the real deal – are also fraying. Thus, what is at stake are the voices of working-class young rappers, the organic intellectuals, who through their music can expose *and* oppose the structural violence woven into the fabric of East London.

Notes

1 The name of the youth club has been pseudonymised as with interlocutors, the youth club's record label, places and any other affiliations with the youth club, to preserve the anonymity of those involved in this research.
2 Kaur, B. (2022) Who Takes the Rap? Young People's Experiences of Violence and Resistance at an East London Youth Club [Unpublished Doctoral Thesis] University of Sussex. This will be referenced throughout.
3 Further discussed in Chapter 3.
4 The term 'open-access' or 'universal' is used to describe services that are openly available to young people as opposed to targeted intervention. However, while

such services tend to be open, they are often located in disadvantaged areas and are therefore targeted in a geographical sense. As stated on the UK Parliament (2011) website, the term 'open-access' is preferred over 'universal'.
5 Due to the sensitive nature of this research, preserving anonymity of the youth club and participants is of significant ethical consideration. In light of this, I have not included the references for Bass Youth Club organisational material, architects, or the housing association that funds the youth club.
6 Rapper of the subgenre Drill.
7 Def Jam Recordings is an American multi-national record label owned by Universal Music Group, based in New York City.
8 A British dating game show.

References

Adams, R. (2019) "Home Sweet Home, That's Where I come From, Where I Got my Knowledge of the Road and the Flow From": Grime Music as an Expression of Identity in Postcolonial. *London Popular Music and Society*, 42(4): 438–455.

Alexander, C.E. (1996) The Art of Being Black: The Creation of Black British Youth Identities. Oxford: Oxford University Press.

Basu, D. (1998) What is Real about 'Keeping It Real'? *Postcolonial Studies*, (1): 371–387.

BBC (2017) PM Vows to Renew the 'British Dream' – but what is it?, *BBC*, 5 Oct 2017, available at: www.bbc.co.uk/news/uk-politics-41506032. Accessed: 14 July 2022.

Berry, S. (2021) London's Youth Service Cuts 2011–2021: A Blighted Generation, available at: www.london.gov.uk/sites/default/files/sian_berry_youth_services_2021_blighted_generation_final.pdf. Accessed: 11 Jan 2022.

Blair, T. (1997) Britain and its Poor. In Tyler, I. (2013) *Revolting Subjects: Social Abjection and Resistance in Neoliberal Britain*. New York: Zed Books Ltd. pp. 153–179.

Bramwell, R. and Butterworth, J. (2020) Beyond the Street: The Institutional Life of Rap. *Popular Music*, (39): 169–186.

Bunyan, P. and Ord, J. (2012) The Neoliberal Policy Context of Youth Work Management. In J. Ord (ed) *Critical Issues in Youth Work Management*. Oxon: Routledge.

Cameron, D. (2008) Britain and its Poor. In Tyler, I. (2013) *Revolting Subjects: Social Abjection and Resistance in Neoliberal Britain*. New York: Zed Books Ltd. pp. 153–179.

Charles, M. (2018) Grime Labour: Grime Politics Articulates New Forms of Cross-Race Working-Class Identities Soundings. *Journal of Politics and Culture*, (68): 40–52.

Cooper, V. and Whyte, D. (2022) Grenfell, Austerity and Institutional Violence. *Sociological Research Online*, (27): 207–216.

Dabrowski, V. (2021) Neoliberal Feminism: Legitimacy the Gendered Moral Project of Austerity. *Sociological Review*, (69): 90–106.

Davies, B. (2019) *Austerity, Youth Policy and the Deconstruction of the Youth Service in England*. Switzerland: Palgrave Macmillan.

Dizzee Rascal (2007) *Dizzee Rascal – Graftin*. Available at: www.youtube.com/watch?v=frN9CoRVWNQ. Accessed: 28 Aug 2022.

Dizzee Rascal (2018) The City and the City. In Hancox, D. (2018) *Inner City Pressure: The Story of Grime*. London: Harper Collins Publishers. pp. 9–30.

Fernandez, R. and Hendrikse, R. (2015) Rich Corporations, Poor Societies: The Financialisation of Apple. Available at: www.somo.nl/wp- content/uploads/2015/10/Rich-corporations-poor-societies.pdf. Accessed: 14 Aug 2022.

Galtung, J. (1969) Violence, Peace, and Peace Research. *Journal of Peace Research*, (6): 167–191.

Ganti, T. (2014) Neoliberalism. *Annual Review of Anthropology*, (43): 89–104.

Garasia, H., Begum-Ali, S. and Farthing, R. (2015) Youth Club is made to get Children off the Streets: Some Young People's Thoughts about Opportunities to be Political in Youth Clubs. *Youth & Policy*, (115): pp. 1–18.

GOV UK (2022) *Government Unveils Levelling Up Plan that will Transform UK*. Available at: www.gov.uk/government/news/government-unveils-levelling-up-plan-that-will-transform-uk#:~:text=For%20decades%2C%20too%20many%20communities,time%20on%20the%20postcode%20lottery (Accessed 13 December 2023).

Gramsci, A. (1971) *Selections from the Prison Notebooks*. London: Lawrence & Wishart.

Hancox, D. (2018) *Inner City Pressure: The Story of Grime*. London: Harper Collins Publishers.

Heather (2012) Do we have an English version of the American Dream? 14 Sep, available at: www.celebyouth.org/the-american-dream/. Accessed: 14 Aug 2022.

HM Government (2022) Levelling Up the United Kingdom. Available at: https://assets.publishing.service.gov.uk/government/uploads/system/uploads/attachment_data/file/1054769/Levelling_Up_the_United_Kingdom__accessible_version_.pdf. Accessed: 2 July 2022.

James, M. (2021) *Sonic Intimacy*. London: Bloomsbury Publishing Inc.

Joseph Rowntree Foundation (2023) *UK Poverty 2023: The Essential Guide to Understanding Poverty in the UK*. Available at: www.jrf.org.uk/sites/default/files/jrf/files-research/uk_poverty_2023_-_the_essential_guide_to_understanding_poverty_in_the_uk_0_0.pdf (Accessed: 13 December 2023).

Littler, J. (2018) *Against Meritocracy: Culture, Power and Myths of Mobility*. Oxon: Routledge.

Made in LDN (2022) *Made in LDN*. Available at: www.london.gov.uk/content/made-ldn#:~:text=Created%20in%202019%2C%20Made%20in,across%20London%20came%20together%20online. Accessed: 14 July 2022.

Merrick, R. (2021) Hundreds of Millions of Pounds of Promised Government Cask for 'Collapsing' Youth Services Shelved. *Independent*, 30 Jan 2021, available at: www.independent.co.uk/news/uk/politics/covid-funding-cut-youth-services-b1791372.html. Accessed: 14 Aug 2022.

O'Hara, M. (2014) *Austerity Bites: A Journey to the Sharp End of Cuts in the UK*. Bristol: Polity Press.

Ord, J. and Davies, B. (2022) Young People, Youth Work & the 'Levelling Up' Policy Agenda. *Local Economy*, (37): 104–117.

Osborne, G. (2012) cited in Dabrowski, V. (2021) Neoliberal Feminism: legitimacy the Gendered Moral Project of Austerity. *Sociological Review*, (69): 90–106.

Oware, M. (2018) *I Got Something to Say: Gender, Race and Social Consciousness in Rap Music*. Switzerland: Palgrave Macmillan.
Perera, J. (2018) The Politics of Generation Grime. *Race & Class*, (60): 82–93.
Perry, I. (2004) *Prophets of the Hood: Politics and Poetics in Hip Hop*. United States of America: Duke University Press.
Randolph, A. (2006) "Don't Hate me because I'm Beautiful": Black Masculinity and Alternative Embodiment in Rap Music. *Race, Gender & Class*, (13): 200–217.
Rose, T. (2008) *The Hip Hop Wars: What We Talk About When We Talk About Hip Hop and Why It Matters*. United States of America: Basic Books.
Rylko-Bauer, B. and Farmer, P. (2016) Structural Violence, Poverty and Social Suffering. In Brady, D. and Burton, M. L. (eds) *The Oxford Handbook of the Social Science of Poverty*. New York: Oxford University Press. pp. 47–74.
Simpson, F. (2021) Spending Review: Sunak's Pledges for Children, Young People and Families. Available at: www.cypnow.co.uk/news/article/spending-review-sunak- s-pledges-for-children-young-people-and-families. Accessed: 14 Aug 2022.
Sköld, D. and Rehn, A. (2007) Makin' It by Keeping it Real: Street Talk, Rap Music, and the Forgotten Entrepreneurship from "the Hood". *Group and Organization Management*, (32): 50–78.
Social Mobility Commission (2020) Monitoring Social Mobility: 2013–2020: Is the Government Delivering on our Recommendations? Available at: https://assets.publishing.service.gov.uk/government/uploads/system/uploads/attachm ent_data/file/891155/Monitoring_report_2013-2020_-Web_version.pdf. Accessed: 14 Aug 2022.
Spohrer, K. (2018) The Problem with 'Raising Aspiration' Strategies: Social Mobility Requires more than Personal Ambitions. 9 May, available at: https://blogs.lse.ac.uk/politicsandpolicy/raising-aspiration-government-strategy/. Accessed: 14 Aug 2022.
Stuart, F. (2020) *Ballad of the Bullet: Gangs, Drill Music, and the Power of Online Infamy*. New Jersey: Princeton University Press.
The Verge (2022) Apple. Available at: www.theverge.com/apple. Accessed: 11 Dec 2022.
Tinchy Stryder (2018) The City and the City. In Hancox, D. (2018) *Inner City Pressure: The Story of Grime*. London: Harper Collins Publishers. pp. 9–30.
Turner, P. (2017) *Hip Hop Versus Rap: The Politics of Droppin' Knowledge*. New York: Routledge.
Tyler, I. (2013) *Revolting Subjects: Social Abjection and Resistance in Neoliberal Britain*. New York: Zed Books Ltd.
UK Parliament (2011) Services for Young People – Education. Available at: https://publications.parliament.uk/pa/cm201012/cmselect/cmeduc/744/74406.htm. Accessed: 27 April 2023.
Unison (2016) A Future at Risk: Cuts in Youth Services. www.unison.org.uk/content/uploads/2016/08/23996.pdf. Accessed: 27 April 2023.
White, J. (2017) *Urban Music and Entrepreneurship: Beats, Rhymes and Young People's Enterprise*. Oxon: Routledge.
White, J. (2020) *Terraformed: Young Black Lives in the Inner City*. London: Repeater Books.
Wiley (2017) *Eskiboy*. London: Windmill Books.

3
~~DANGEROUS~~ DADA?
Reconceptualising UK Drill as Avant-Garde

Wanda Canton

By comparing the popular rap form, UK Drill (UKD)[1] to post-war Dada art, UKD is reconceptualised as an artistic expression instead of a criminal one. My purpose is to offer a political intervention rather than an artistic critique, to respond to the urgent problem of criminalising music, particularly as it censors and sanctions Black artists. I begin by contextualising UK Drill and Dada in their respective socio-geographic origins. There are shared artistic styles respective to each movement, although neither promotes a specific ideological doctrine. Nevertheless, both Drillers[2] and Dadaists have offered political commentary and have been used for political purposes, including to their detriment in the case of Drill. I move to consider the Dada *ready-made* technique of using everyday objects as templates, comparing this to UKD's simplified rhythmic structures, which democratises musical participation to some degree. The literary/lyrical content of both Drill and Dada includes violent and aggressive language, which I argue reflects violent society rather than causes it. This includes institutional violence, which criminalises Drill and contributes to the perceived need for Drillers to mask themselves with balaclavas to conceal their identity. Simultaneously it enables audiences to imagine themselves 'Behind the Bally,'[3] as my subheading refers. Dada also used masks to distort and disturb, problematically relying on representations of Blackness to do so. Finally, I turn to the most notable distinction between the two, namely, that whilst Dadaism is championed as an iconic disruption of visual and literary arts, Drillers are continually obstructed from public space. I suggest this is in part due to its creators being racialised.

Although there is no scope here to discuss in detail whether the current knife crime epidemic is statistically accurate or not, the general perception is that it is rapidly rising. Mainstream media made only one mention of

DOI: 10.4324/9781003361046-4

knife crime in 2000 but by 2008 there were over 2,000 references, creating a widely held belief that it is a new and escalating problem (Elliott-Cooper, 2021). Researchers have stated that knife crime is 'one of the most used and least understood crime labels in popular parlance' (Williams and Squires, 2021: 13), and there have been many changes to the way in which knife crime is recorded; producing new, previously unavailable data. Reports have claimed there are 11 serious stabbings a day and the 20% hike in murders between 2019–2020 has been attributed to a 107% increase in gang affiliation, although there is no clear definition of gangs (Harper, 2017; Griffiths, 2021). Both national and local policies cite children and teenagers as the predominant victims to the extent it has become a 'public health crisis' (Brighton and Hove, 2020; APPG, 2019). But there are concerns that they are also increasingly the perpetrators of violence. This has led to material consequences for young people and scrutiny of their popular practices including music, for fear it incites violence. UKD has become the centre point of debates around community safety, knife violence, and policing. Many Drillers have been implicated in UK street violence as both victims and perpetrators, sometimes both, such as Incognito who was murdered a short time after being accused of fatally stabbing someone else (BBC, 2018). UKD has repeatedly been held responsible not just for glorifying violence but directly causing it, leading to numerous attempts to deter artists through restrictive orders, suspended prison sentences, and censorship to criminalise the performance or distribution of Drill, including on social media platforms (BBC, 2021; Fatsis, 2019; Rahim, 2019). Increasingly, the music itself is criminalised rather than specific, actual acts of violence. Given that the predominant creators of Drill and rap more generally are young, Black (African-Caribbean) men, it is concerning that this demographic is continually conflated with criminality. It is therefore urgent to consider the repercussions of criminalisation and how it polices who is and is not present in public, cultural spaces. To be clear, this is not to justify violence nor to dismiss its real devastation and impact. Unfortunately, there have been and continue to be many brutal attacks and the traumatic aftermath is felt by the family, friends, and wider community of those involved. Every life taken is a loss. This chapter focuses on the music that has been blamed for street violence in the UK context. It also considers whether existing policing responses utilise discriminatory and violent practices, which are unlikely to resolve the tensions and inequalities in the communities most heavily policed.

Drill vs Dada: A Brief History

In 1916, two poets, Hugo Ball and Emmy Hennings approached the Meirei bar in Zurich and proposed to host a literary cabaret. This would create a space in neutral Switzerland for artists and those fleeing WWI[4] to come together to share music and poetry. *Cabaret Voltaire* was a watershed moment in the birth of Dadaism. Although early Dada had a literary focus, it would go on to embrace a range of art mediums including sculpture, collage, theatre, and dance.[5] Dadaism grew in Berlin, Paris, and New York, among other cities, with localised emphasis on certain styles or forms. Dada continues to inspire artists today, given its focus on state violence and war, such as Adam Pendleton's 2018 *Black Dada Flag (Black Lives Matter)* for Frieze New York (Buckley and Tsai, 2020; Haakenson, 2019). Speculation about the meaning of the name 'Dada,' includes potential French/Italian/Russian etymologies or having no meaning at all (Tzara, 1918; Ball, 1916a). Others have suggested it represents the repetitive percussion of Dada poetry, noting its first appearance in 1916 as part of the poetry-based *Cabaret Voltaire* (Garner, 2007). There is some reluctance to define Dadaism as a movement per se given its focus on form rather than sharing a particular doctrine. It has been argued that 'Dada's importance was as a state of mind, not as a style or movement' (Arnason, 1969: 307). I will nevertheless refer to it as such, primarily for purposes of clarity.

Aside from the many shared sentiments among Dadaists, notably anti-war and anti-art, Dadaism was a commitment to the avant-garde. This refers to innovative and experimental cultural or artistic work, deliberately pushing the boundaries of existing practices to find new forms and ideas. Rather than conforming to a specific principle or form, Dada shared an artistic ethic (Richter, 1964). This included an interest in spontaneity and everyday materials. Although not proscriptive or dogmatic, there are commonalities in Dadaist styles resulting from an interest in chaos and disruption, as shall be explored. The context of violence within which Dadaism emerged united individual artists motivated by a critical, anti-war position. Many Dadaists were already professional working artists and had engaged with other techniques, such as Marcel Duchamp whose earliest works followed the Impressionist tradition, but by the 1920s he had become a central member of the Parisian Dada scene. Dada has been compared to the earlier Italian Futurist movement and the later Surrealism but remains distinct due to its socio-political commentary against war and bourgeoisie culture (Garner, 2007). Several Dadaist manifestos reveal these intentions, as will be discussed.

UK Drill shares similar attributes to Dada in so far as it developed an artistic intent or ethic, without a cohesive ideology and with localised differences. Chicago rapper, Chief Keef is credited as the pioneer of Drill with his track, *I Don't Like* in 2012. Like the international splintering of Dada, Drill has developed in numerous countries with regional styles and slang, the breadth of which is not discussed here. In the UK, Drillers favour an aesthetic of balaclavas and masks, references to knives and 'ops' (opponents). UK producers developed instrumentals using unique snare patterns, hi-hats, and 808 slides, which create a deep haunting bass and mark a distinctive UK Drill sound. Individual or group monikers and song titles show affiliation with specific inner-city boroughs such as OFB (Broadwater Farm, Tottenham), Unknown T's song 'Homerton B' (a region of East London), Moscow17 (South London postcode) and Zone 2 (referencing Peckham). Drillers rap at slower beats per minute (BPM) than earlier UK rap genres such as Grime, as will be demonstrated. Like 'Dada,' there is no clear consensus on the meaning of 'Drill.' Some have claimed it represents a drilling or whirring sound in the instrumentals, but as this is not particularly discernible, it is less convincing than the suggestion that drilling is a slang term for automatic weapons (Fatsis, 2019; Lee, 2022). Drillers do not adhere to any consistent political message.[6] Subgenres range from Gospel Drill with religious incentives, to politically motivated Drill artists such as DrillMinister who sought to run for London Mayor. Despite the variety and ever-changing nature of Drill, there does tend to be a shared distaste and mistrust of the police, common to other forms of rap (Canton, 2022).

To introduce the comparisons I make between Drill and Dada, I have below created a visualisation of an excerpt from Hans Richter's (1964: 9) written explanation of Dada. I have replaced the word 'Dada' with 'Drill' or 'Driller' to demonstrate how Dadaist manifestos could be applied to Drill. I have reproduced this in the style of a Dada poster/poem by using a visual dislocation of language and collage-like style (Figure 3.1).

Ready-made Dada, Imitable Drill

The Dadaist 'ready-made' technique created new objects from existing materials and items. Tools and appliances from everyday life, mundane in their familiar use, were distorted when welded to other objects. Man Ray's *Gift* (1921) featured an iron with thumbtacks glued to its sole, creating spikes and turning the iron from a household appliance into a weapon. The ready-made offered commentary on the absurdity of war; the unremarkable turned deadly, the tedious turned terrifying, that ordinary life can be so suddenly and completely disturbed. Dada's abstract art advocated dissolution, anarchy, championing a philosophy and practice of anti-art and the impossibility of originality. Dada mocked the high-brow art world by quite literally, pissing on it. In 1917, Marcel Duchamp sought to exhibit an upside-down urinal and called it *Fountain*. Although the original was never exhibited, it would go on to become one of the most well-known examples of ready-made Dadaism, reproduced in multiple replicas. When asked whether he had purposefully contributed to the 'discrediting' of art, Duchamp (1966) remarked; 'deliberately, yes … I really want to get rid of it [art]. There's sort of an unnecessary adoration of art, today.' Hugo Ball noted that mocking

drill was not an artistic movement in the accepted sense; it was a storm that broke over the world of art as the war did over the nations.

It came without warning, out of a heavy, brooding sky, and left behind it a new day in which the stored-up energies released by

drill [was] evidenced in new forms, new materials, new ideas, new directions, new people – and in which they addressed themselves to new people.

drill had no unified formal characteristics as have other styles. But it did have a new artistic ethic from which, in unforeseen ways, new means of expression emerged.

These took different forms in different countries and with different artists, according to the temperament, antecedents and artistic ability of the individual **driller** The new ethic took sometimes a positive, sometimes a negative form,

often appearing as art and then again as the negation of art, at times deeply moral and at other times totally amoral.

FIGURE 3.1   A visual except of Richter's (1964) description of Dada replacing 'Dada' with 'Drill.' By Wanda Canton.

and parodying classical artistic concepts sought to challenge the privileging of Europe as an intellectual, cultural leader, which disavowed its violent conquests:

> Our cabaret is a gesture. Every word that is spoken and sung here says at least this one thing: that this humiliating age has not succeeded in winning our respect. What could be respectable and impressive about it? Its cannons? Our big drum drowns them. Its idealism? That has long been a laughingstock, in its popular and its academic edition. The grandiose slaughters and cannibalistic exploits? Our spontaneous foolishness and our enthusiasm for illusion will destroy them.
>
> *(Ball, 1916b: 61)*

On one hand, it may have been that the Dadaists sought to diversify their membership (or following) by creating open cabarets and opportunities to participate through the ready-made technique, given that it is more easily reproduced and imitable. On the other, many of its most celebrated members were part of the same exclusive world and networks they were criticising. Duchamp's obscene sculpture, initially submitted under the pseudonym, Richard Mutt, may not have been considered worthy of analysis and replication without Duchamp's existing prominence and credibility as a painter. Nevertheless, Dada works were created as forms of spontaneous curation rather than being based on the technical skills of painters, and everyday materials provided a template from which something new could be made. Theoretically, anyone can participate in its production and therefore the hierarchy as to who and what can be considered artist/art was being challenged.

The 'chance' poetry of Tristan Tzara and Dadaist poetry in general could be considered the literary form of the ready-made. According to Tzara (1918), 'the new artist protests: he no longer paints.' Distorting and weaponising common objects included an attack on language. Tzara would cut up words and sentences, arranging them into a stanza (verse) as they fell haphazardly after being placed and plucked from a jumbled bag. The ready-made was incorporated into 'simultaneous poetry,' which utilised sounds of sirens, crashes, and everyday noise (Ball, 1916b). This could be compared to UKD's use of city sounds from sirens, gunshots, motor revving, and barking dogs, adding a layer of phonetic aggression to the lyrical content. Hugo Ball's incoherent poetry such as *O Gadji Beri Bimba* in 1916 'almost caused a full-blown riot' (Arnason, 1969: 307). His repeated nonsensical sentences evoke a sound akin to a prayer disintegrating into African-esque chanting (Aurbacher, Maier, and Liska, 2015). He named these forms *Verse ohne Worte* (poems without words), and *Lautgedichte* (sound poems) (Ball, 1916b: 70). Other works such as *Karawane* (2020) sound in parts like Germanic or French instructions, which may have been in reference to the war. From its early formation to contemporary developments, Dada has sought to reflect the destruction of socio-political environments in the chaos and violence of its form, be it sculpture or literature. Dada leaned towards Nietzschean ideas of self-destruction as necessary for moral cleansing, whereby future peace could be achieved only through the complete annihilation of the violent present. It claimed to be 'a reaction to the general disintegration around us' (Richter, 1964: 48). Hugo Ball passionately declared 'they cannot persuade us to enjoy eating the rotten pie of human flesh that they present to us. They cannot force our quivering nostrils to admire the smell of corpses' (1916b: 67). Such sentiments would continue throughout the development of Dadaist poetry, including Amiri Baraka's 1965 *Black Dada Nihilismus*, written in the same year as the assassination of Martin Luther King, reviving the context of violence from which Dadaism speaks. It begins by asserting 'poems are bullshit' and calls for there to be 'no love poems written until love can exist freely and cleanly.' This reflects the consistent rejection of the aesthetic by Dadaists who could not abide superficial celebrations of arts and culture in a world brutalised by war.

Elements of the ready-made can arguably be found in rap music in the use of sampling, which, according to iconic rap group, Public Enemy, enables 'brand new creation made out of found objects' (BBC, 2011). Across different genres of rap, from US hip-hop to UK Grime, there are clips and references to sounds in city environments, such as sirens, cars, dogs, which represent the use of everyday materials. Even the primary instrument of rap, the voice, could be considered as an immediately available resource, without relying on buying and training in classical music or instruments. This expands the possibility of who can make music and due to rap's commercial success, non-classically trained rappers hold their own among more conventional musicians. Although there is no scope here to outline a full history of rap, or indeed its relationship to other music forms, the strong consistency of sampling, environmental sounds and use of the voice suggests rap has long made use of the ready-made technique.

UKD uses sampling alongside simplified rhythmic schemes and slower BPM. The earlier rap form, UK Grime, delivers fast flow, multisyllabic constructions, and consistent rhythmic schemes. Drill rap, on the other hand, is around half the BPM of Grime and does not contain the complex assonance, consonance, or syllabic structure. There are exceptions to this, particularly as Grime artists increasingly utilise Drill instrumentals. However, UKD provides a general template of basic rhythmic patterns, which can be more easily replicated by prospective artists. Simplifying the rap expands the membership of who can become a Driller. No longer premised on who *spits* (raps) the fastest, the most complex, or most poetic, Drill offers a consistent structure, which broadens the scope of who can 'do' Drill, through imitation and innovation. Early UKD producer, MKThePlug, acknowledges that initial Drill beats sounded similar in part because pioneers in the scene were learning from each other how to produce Drill instrumentals (Keith, 2020). This created a consistent foundation upon which new artists could add, develop, and introduce new changes and trends. Although I am focusing on the linguistic-visual components of UKD, further work might consider whether early Drill instrumentals were intended to create a template, thus ready-made, for others to replicate. For example, Grime artist, English Frank's (2011) SBTV *Warm Up Sessions* uses an imperfect AAAA[7] with multisyllabic constructions throughout, compared to Drill group, R6 who use a less consistent ABCB rhyme scheme in *Redrum* (R6, 2016). Whereas English Frank might use an average of 12 syllables to each bar, R6 use six. Like Dada, Drill sometimes appears to use nonsensical lyrics to construct rhyme or repetition to provide syllabic consistency. This can be heard in R6's (2016) repetition of the word, money.

Since Dada, the exhibition space for contemporary arts and culture has changed significantly. No longer constrained to galleries and art houses, social media provides an accessible platform upon which anyone can upload music. For example, the former Channel U and current YouTube channels such as LinkUp TV, SBTV, and GRMDaily have enabled new artists to 'cross over' into the mainstream. Stormzy, who first appeared on SBTV in 2010 became the first Grime artist to headline Glastonbury festival in 2019. To some degree, the development of mass online followings has created 'ready-made' audiences for new genres to tap into. New artists have the potential to reach global audiences instantly without necessarily relying on specialist networks. Krept and Kronan's second ever video went viral, for example, generating massive publicity for the duo who would go on to become some of UK rap's most recognisable names. Their video, *Otis*, was shot in a car/park with minimal editing. Similarly, Drillers Digga D and ArrDee produced a video for their song *Wasted*, which cost 'tens of pounds' (Thapar, 2021). Consistent with Dadaists disdain for the exclusivity of art institutions, social media expands the possibility for others to engage with and reproduce artistic trends. Theoretically, this democratises public or cultural space, as less experienced artists can (in principle, at least), distribute their music alongside professional and high-profile artists. However, as the Driller populace increases, artists must find alternative means to stand apart. This saturation of content leads to an attention economy whereby users must continuously find ways to generate views on such competitive forums, leading to more risky and shocking content being created (Stuart, 2020). Aside from the accessibility of user-generated content, social media popularity is partly premised on imitability, whereby 'trends' start with different individuals or groups creating the same content. Examples include challenges, dances, and memes (see Chapter 4). For example, rap challenges ask different members of the public to recite the same Drill lyrics (King, 2019). The recent *You Don't Look British* make-up trend involves predominantly women styling themselves as 'chavs' whilst lip-syncing the lyrics from a widely ridiculed Grime song from Blackpool (Dancemyday, 2022). This easily imitated presence on free, global platforms, and the conscious attempts to 'go viral' may contribute to the fear of Drill being contagiously influential.

Dada as Reflection, Drill as Cause

Some have criticised the accessibility of social media, claiming it popularises and promotes the perceived violent content of Drill. From the New York Mayor to British parliamentarians, politicians around the world have 'declared war' on Drill and seek to have it removed from social media platforms (Vozick-Levinson, 2022). Lyn Brown, Labour MP, conflated Drill with the deliberate, abusive grooming of children:

> The murders [which] this video is about may be fictitious, but by looking at the online comments we quickly see many young people who believe it is real ... The law may be unclear about whether such videos illegally incite violence, but I believe they are dangerous. They make the grooming of children easier by glamorising drug dealing and murder as a lucrative and exciting alternative to the hard and unrewarding work they see demonstrated in the lives of their parents. Presented as an alternative economic model, it is offered to children and made to look exciting.
>
> *(Brown in UK Parliament, 2018)*

To address these claims, firstly it assumes that young people are not able to distinguish between what is real or not in Drill, despite likely consuming a variety of violent content in films, video games, or news, which may or may not be fictitious. There is a further assumption that social media comments accurately reflect the way in which UKD manifests in the day to day lives of listeners, if at all. At the time of writing, I reviewed the comments for the song, *He's Dead* by Moscow 17 on a YouTube upload which has since been removed. The track was released as a response to the notorious *No Censor*, originally released in 2019 and reuploaded in 2022 following its removal from YouTube after only five hours of publication for incitement of violence (Moore, 2019). Both songs are notorious for listing the names of individuals who have been murdered. I scrolled through 72 original YouTube comments (excluding replies) on *No Censor* and hundreds of the newest comments on *He's Dead* to consider the reactions to it. None of the comments in English at least made any reference to an intention to harm other people, the vast majority praising the beat or flow of the track and several leaving more critical comments condemning the lyrics. There is little evidence or explanation as to how Drill incites listeners to enact violence other than it being 'cool.' There are claims that the predominant audience of UKD[8] are White teenagers, as stated by Drillers themselves (Trend Centrl, 2020). Yet the predominant rhetoric around knife crime alludes to it being a problem in Black communities, disproportionately constituting victims and perpetrators (ONS, 2020). If correlations between listening and enacting violence were so concrete, one would therefore expect knife violence to be an issue affecting impressionable White boys. This is not consistent with the general perception.

Gangs, which existed well before Drill, do operate in the UK, as they do all over the world. They are strengthened where there is a lack of stability and opportunity, making them a viable or perceived necessary alternative economic revenue as Brown acknowledges. Therefore, whilst they can certainly be violent and exploitative, including of children, they do not exist in a vacuum without cause and are a symptom of wider socio-economic problems, which enable them to thrive (Lynes, Kelly, and Kelly, 2020). Failing to acknowledge the structural causes of gangs falls danger to insinuating that crime and violence are inherent to particular demographics (namely, Black communities), and blamed on cultural practices (namely, rap). Not only does such a reductionist analysis reinforce tropes of racialised people as intrinsically violent, it provides no real solutions for addressing and preventing crime, other than the absolute control and segregation of such perceived *Natural Born Killaz*.[9] If violence is innate rather than structurally caused, outlawing Drill would be ineffective as such personalities would persist irrespective. Some criticise Drill for commodifying stereotypes of road life to build capital, by reproducing caricatures to amass larger followings (Stuart, 2020; Trend Centrl, 2020; Canton, 2022). By this argument, Drillers profit from tropes of gang culture without direct experience of it and consolidate associations of rap to criminality. Whether or not Drillers are actual gang members themselves or have direct experience of the stories in their lyrics, they certainly do face over-policing, surveillance, and scrutiny. As predominantly young, Black men, they are already subject to gang rhetoric regardless of actual involvement. Prospective financial gain is matched by losses to freedom and opportunities.

Nonetheless, Drill does contain violent lyrics, and this is inescapable. However, the conflation of rappers with gangsters leaves no opportunity to distinguish between actual and symbolic violence. UKD is known for referring to or threatening retaliation against rivals. The assumption that 'ops' always relates to gangs negates the possibility that it might refer to rival artists who are competed against. Rap 'battles' or 'clashes' have long staged competitions between MCs artistically but, despite the language of conflict, these are not literal or physical fights. Drill videos sometimes explicitly display weapons but more often include gestures. Some symbolic references or miming of stabbing or shooting are present in dance forms. For example, finger guns can be seen in the dancehall *bogle* motif popularised by major artists including Rihanna, who has not been accused of gang affiliation. Common Drill dance styles include the *Moscow march* imitating the revving of a vehicle, *skengbop* (literally 'knife dance'), and *kick hops*, which look like stamping on or kicking something. Removing these gesticulations from their artistic-cultural context leaves them vulnerable to misinterpretation. They become more problematic when they are included in music videos mocking or listing individuals who have been attacked or killed in real life, which does nothing for the protestations of Drillers' claims that their lyrics/gestures are fictional. However, if Drillers committed all the murders referenced in their songs, UK cities would seemingly be overrun by serial killers. In *Secrets Not Safe* by Chinx (2022), there are references to around seven attacks. By my count and interpretation, *No Censor* by Zone 2 (2019) names around 25 people who have been killed, attacked, or threatened. Loski and other Drillers have consistently said that lyrics are not real confessions or self-snitching,[10] but are intended as entertainment. Further, despite representing 'road life,' (street life, hustling) artists state they are consciously seeking and providing a positive 'off road' (legitimate) alternative (Trend Centrl, 2020). There is frustration among some rappers who feel the police and society want them off the street but won't allow them to pursue the (music) career enabling them to do that (Lee, 2022). Even if exaggerated, UKD may be a performance of the violent environments in which Drillers live. Yet it is frequently conflated with creating them.

Dada was also known for its violent language and violence *to* language as already discussed in the earlier example of Hugo Ball (1917a) whose nonsensical chanting over discordant piano and evocation of militaristic instructions created a disconcerting attack on the poetic-phonetic form. Other examples include graphic language in manifestos including Tzara's (1918), with vocabulary such as: *piss, flabby, insipid flesh, torture, carnage, corpse-flowers, manure, shit, disease, diarrhoea, incest, entrails, putrid rats infecting the bowels of the bourgeoisie, decomposition, gonorrhoea of a putrid sun*. Tzara used whistles, screams, and noises throughout his performances and many poets used aggressive noises and expletives as part of their piece (Richter, 1964). Dada deliberately sought to provoke rage and disgust in its audience, claiming that:

> everything must be pulled apart, not a screw left in its customary place, the screw-holes wrenched out of shape, the screw, like man himself, set on its way towards new functions … Until then: riot, destruction, defiance, confusion.
>
> *(Richter, 1964: 48)*

Dada poets sought to deliberately 'agitate the crowd by attacking poetry' (Grindon, 2011: 90) and artists such as Huelsenbeck insulted audiences directly (Jones, 2014). Violence in Dada was not limited to metaphorical representations but was physically enacted. In an artistic context this included provoking aggression in audiences such as Man Ray's *Object to Be Destroyed*, which was indeed destroyed by gallery visitors. In another example, an effigy of a German soldier was hung from the ceiling of the First International Dada Fair in Berlin, 1920 (Richter, 1964). This evokes parallels to the visuals in Skengdo, AM and DrillMinister's video, which shows a bloodied policeman[11] (MixTape Madness, 2019). There have been claims of interpersonal violence between Dadaists, such as Emmy Hennings and Hugo Ball, and a rumoured pistol duel between Tzara and Arp (Richter, 1964). Yet Dadaism was not subject to the same scrutiny levelled at Drill. It was credited with protesting social violence even by embodying and provoking violence in others.

Arguably, Dada linguistic violence is an example of Walter Benjamin's (1921) *divine violence*, which annihilates law and destroys boundaries for revolutionary aims. In Dadaism, the laws and boundaries of language and sound. Divine violence is utilised for radical (or more specifically to Dada, anti-war) intentions but ceases to be divine or just, once it re-establishes laws and boundaries, albeit new ones. If revolutionary violence becomes state-sanctioned, it is no longer revolutionary. If the avant-garde becomes conventional, it is no longer avant-garde. Therefore, violence, which remains in an artistic context, has a particular subversive ability to agitate and provoke without being reinstated into the socio-political authorities it protests, given that art has no legal authority. There was no intention for Dada appliances-stroke-weapons to become actual weapons, no serious claim to destroy linguistic rules in favour of gibberish. Therefore, there is no reconstitution of laws or boundaries, maintaining its divine quality. This would be consistent with the name 'Dada' having no meaning, nor intention to produce new vocabulary or linguistic standards. It is in the peripheral nonsense that Dada sought to best serve its anti-war agenda: making a mockery of violence by amplifying it. This alludes to a wider question present in both the case of Dada and Drill, that is, can art imitate violence without reproducing it?

Man's Talking about War[12]

Although it would be crass to compare it to a world war, UKD has emerged within a war of sorts. In mainstream discourse, this war has been posited as a war between peers, given the complicity and victimisation of some Drillers in street violence. This has been from its outset, given the founder of Chicago Drill, Chief Keef was implicated in the murder of Joseph Coleman. Rivalry between rappers has associated subgenres with violence since the East Coast/West Coast 'beef' in 1980s American hip-hop, wherein the murders of Tupac Shakur and Biggie Smalls were blamed on gang fighting between the Bloods and Crips (Lynes, Kelly, and Kelly, 2020). However, the notion of war could also be considered as one between the police and racialised communities. It would be naïve to suggest that inner-city boroughs have the same arsenal as state institutions, so it is not a war of equal nations. However, the Metropolitan Police declared *war* on knife violence in 2008 followed by Prime Minister David Cameron's 'all-out war on gangs and gang culture' in 2011 (Elliott-Cooper, 2021). This has led to a variety of policing practices, which label people of colour as 'suspect communities,' disproportionately suspecting or outright accusing them of affiliating with gangs, most notably African-Caribbean boys/men. Some have compared the over-presence of police in boroughs like Tottenham to military occupation, demonstrating the extent to which this has been experienced as a form of warfare (Elliott-Cooper, 2021). At its most extreme, police warfare has resulted in the fatal shootings of, predominantly, Black men. This includes Mark Duggan in 2011 who was widely portrayed in the media as a gangster, using his image to depict a hard-faced thug whilst omitting the context of the photo being at his daughter's graveside (Aitkenhead, 2014). The police officer who shot Duggan claimed he believed there was an imminent danger to his life (Dodd, 2019). More recently was the shooting of Chris Kaba in September 2022 who was also a Driller. He was killed after police, in an unidentified car without sirens, instigated the same 'hard stop' technique strongly criticised after Duggan's killing, whereby a vehicle is violently stopped and blocked by armed officers (IOPC, 2022). Given that Kaba was shot in the head through his front windscreen, it is difficult to comprehend how he could have posed an immediate threat to the life of the policeman who killed him. The rhetoric of war has been used to justify legislation and taskforces, which, to name a few, have included:

- 2005: the use of risk assessment policies, such as Form 696, to obstruct events involving DJs and rap music as 'high risk,' therefore disproportionately targeting Black-created music,[13] and cancelling concerts.
- 2011: the Trident Gang Crime Command, which focused specifically on gun crime and homicide within the Black community,[14] is responsible for the shooting of Mark Duggan.
- 2012: the Metropolitan Police Gang Matrix is established, which registers gang affiliation according to social media use and *music* preferences. By 2018, 70% of those registered were young, Black men and repercussions of registration (without conviction) have led to evictions, loss of education/employment and the harassment of bereaved families.[15]
- 2013: an increase of Joint Enterprise[16] convictions, a third of which were Black people despite representing 3.3% of the population at the time. Rap videos were used as 'evidence' of gang membership (Elliott-Cooper, 2021).
- 2016/2017: Black people were eight times more likely to be stopped by police and three times more likely to be arrested than White people.[17]

It is significant that UKD did not arrive in the UK until 2014, and by 2016 it had been associated with knife crime. This was the same year that UK Drillers had started to gain popularity via breakout artists like Loski and R6 (Keith, 2020). Therefore, Drill developed in the wake of severe policing policies and was quickly associated with gangs and knives even though the 'epidemic' of knife violence had been ranked as the number one policing priority since 2008, higher than terrorism (Williams and Squires, 2021). This was six years before British producers began making Drill and four years before Chicago Drill arrived.

To take the single case of Driller Digga D, he was first arrested whilst in school, imprisoned numerous times where he was stabbed in the eye and held in segregation, received a Criminal Behavioural Order demanding he report his music to police and, after being released from recall following a breach of his probation licence, he was stopped and searched by police within five days and then again three hours later, all captured and documented on the 2020 BBC documentary, *Defending Digga D*. In the documentary, he casually mentions suffering from Post-Traumatic Stress Disorder and is relentlessly harassed and monitored by police. References in Drill lyrics commonly refer to 'jakes' (police), 'obs' (surveillance), 'catching a case' (being charged with an offence), and being incarcerated. This is unsurprising given that Drill has been criminalised under a variety of laws relating to gangs and terrorism, which has made the music itself a crime rather than specific or subsequent acts (Fatsis, 2022; 2021; 2019). It has been conflated with hate crime and blamed for 37% of gang homicides with music groups such as OFB being called a gang without much explanation or definition (Falkner, 2021). Campaigns are ongoing to censor or outright ban Drill from social media and streaming platforms. Drillers have even been restricted from wearing particular items of clothing or speaking certain words publicly (Elliott-Cooper, 2021). This also means that creating and sharing music has itself become a form of resistance, as will be discussed in the final section. This is significant in noting that irrespective of the precise political intentions of Drillers, the very creation and engagement with it is political.

Drillers have claimed that their music reflects the extremity of their everyday environments, which, as the above demonstrates, may include institutional violence. As DrillMinister put it, 'if I grew up in Cambridge or grew up in Shropshire … I'd be talking about rowing boats' (LBC, 2020). It is widely accepted by rap scholars and supporters that rap such as hip-hop is social commentary, whereby rappers observe the world around them and express it through music (Rose, 1994; Keyes, 1996). This informs the lyrical content of UKD as well as the instrumental choices and specific use of rap as percussion. Bkay, a pioneering producer of UKD, explains: 'we wanted something from the gutter' (Keith, 2020). This suggests a deliberate intent to create a dark or aggressive sound, which is not unlike the brutalism of Dada, to mirror the violence of war and society through its aesthetic and language. This is reiterated by Driller, LD who states 'I'm really from the trenches … I'm from the gutter, but you learn a lot of lessons there' (Noisey, 2022). To reflect the realities of young, British people in UK inner cities, the instrumental and lyrics correspond. For early hip-hop (and indeed, Dada), there was 'no reason to sing sweet harmonies' during the mundanity of unemployment and oppressive city life, so the style of repetitive, percussive rap reflected this (Chang, 2007: 48).

Like Dada, Drill provokes a reaction in its audiences, namely, outrage. Consider the applicability to Drill in the following quote:

> To outrage public opinion was a basic principle of Dada ... The devising and raising of public hell ... And when the public (like insects or bacteria[18]) had developed immunity to one kind of poison, we had to think of another.
> *(Richter, 1964: 66)*

Although I want to avoid terms such as 'incitement' given its criminal connotations, both mediums are deliberately provocative. For Drillers, this may be less a conscious move, rather informed by the attention economy I have already referred to, whereby each new track and artist must supersede what has gone before. Whilst Dadaism had an explicitly political undertone to this approach, Drillers are not necessarily activists per se. However, the repercussions of the genre may nevertheless have political outcomes, as will later be explored. Both Dada and Drill, whether intentionally or not, stimulate strong, sometimes violent, reactions in their audiences. Significantly, in the case of Drill, these violent reactions are not necessarily in their anticipated young audience, but rather in the institutions which oppose it. These reactions become a staging of society itself as institutions literally enact and perform the very behaviours being criticised. For example, the fierce crack-down and policing of Drill legitimises the references to over-policing and surveillance in the music. This would be consistent with Attali's (1985) description of music as mirroring society, suggesting that UKD reflects violence and tensions with the police, rather than causes it. What is pertinent about the analogy of a mirror reflection is the element of recognising oneself in what one sees. On the one hand, audiences may relate to realities depicted in UKD. That is, they see a reflection of their own lives or experiences. On the other, it implicates the audience who, confronted with the performance of themes, such as police brutality, racial discrimination, and social disenfranchisement, is forced to consider whether such representations reflect their own privilege, power, or prejudices. This should be considered in an institutional context (the police) as much as the individual. Rather than engaging with such unbearable accusations, they may be inclined to rubbish and outlaw UKD, thus inducing more critical commentary, more counter-resistance, and so the cycle repeats.

Behind the Bally: Facial Awareness

Having outlined the constant threat to outlaw Drill music, the traditional use of balaclavas or 'ballys' by its artists can be understood primarily as an anonymising strategy. Many Drill videos feature artists and even models in balaclavas or ski-masks. Variations can include Covid-19 medical masks, 'viper' mouth masks, or tied bandanas. Examples include LD, Zone 2,

DrillMinister, AM, Buni who wears a Venetian carnival style mask, and M Huncho, who wears a metallic-like, hockey mask. He is not a Driller per se, rather more aligned to Trap music, but collaborates with Drillers and can be said to be associated with and inspired by the scene. Artists and their entourages often match masks with dark, baggy, non-distinct clothing, and gloves (keep this masking of the body in mind for a later Dada parallel). Drillers have cited reasons for masking as protecting their identity from the police and rivals, relatives, teachers, and community figures more widely (Thapar, 2019). Facial recognition software is increasingly used by authorities and, notably, was first trialled at London's Notting Hill Carnival, which, as a celebration of Caribbean culture, targeted racialised groups (Elliott-Cooper, 2021). It is therefore unsurprising that Black artists feel disproportionately at risk of surveillance. When anonymity is protected one may also feel more at liberty to say or do things without facing the same repercussions as they would exposed. Some Drillers have alluded to the balaclava as facilitating an alter-ego whereby their anonymity enables a different or heightened personality to manifest (Thapar, 2019). It may further enable them to build a following through intrigue, effectively acting as a trademark or branding. This may, however, have the reverse effect, as there is added interest in identifying artists on the part of the police, to hold them accountable for offensive language and to make prosecution possible. It is not always effective in anonymising the artist in this regard.

Masking (partially) conceals identifiable characteristics such as ethnicity, age, or personal attributes, which could reduce the relatability of artists to audiences. For example, listeners might respond more to musicians who look, sound, and have lived/do live like them. Typically, we especially emulate or admire public figures if they represent our own identity and experiences. 'Near Peer Role Models' are inspiring because they are like us in some way and represent the potential for excellence (Murphey, 1998). Effectively, their success signals that it can be obtained by others who share similarities. This is not limited to rap music and is demonstrable in genres such as rock-folk, whereby music reflects the experience of a community (Frith, 1981). I propose however, that the absence of the artist's identity can conversely widen the scope of audience identification. The lack of a specific identity liberates the audience and invites a wider listenership to imagine their own relatability to the anonymous artist. You do not have to conform to an identifiable group if the identifiable group is hidden. Cavarero (2009) notes that the face is the most unique and human attribute of the body. Therefore, by concealing facial features, even if it does not entirely mask the artist's other identifiable traits, including ethnicity, the viewer is free to imagine whose face is beneath the bally, even their own. The use of balaclavas does not limit the audience's imagination by the artist's actual identity, effectively creating a blank canvas upon which they can project their own associations or fantasies. M Huncho

(2020) observes that aside from providing him privacy, anybody could wear his trademark mask and claim to be him. This returns to the earlier point of imitability that widens the participation of prospective Drillers who, in *ballying up*, can claim membership to the community. Although it is not the focus of this chapter, further discussion might consider whether this facilitates White listeners (the supposed majority consumers of UKD) access to cultures and genres of rap music. However, such absence also exacerbates opportunities for critics to project phobic imaginations, such as one-dimensional archetypes of gangsters devoid of individual personalities, nuance, or histories. Given that balaclavas are associated with crime, such as robbery, a double association of rap to Blackness and balaclavas to crime may consolidate phobic associations in some audiences. I am reminded in both instances of the quote attributed to writer, Anaïs Nin, that we see things not as they are, but as *we are*.

Dada also distorted the face and body to induce and represent the absence of individuality, violently enforced in war. Hugo Ball used full-body masks to create oblong, cylinder silhouettes, which restricted individual expression.

> A world of abstract demons swallowed the individual utterance, ingested individual faces into masks tall as towers, swallowed private expression, robbed single things of their names, destroyed the ego, and shook out great colliding seas of chaotic and confused feelings.
>
> *(Ball, 1917b: 225)*

Cavarero (2009) refers to *horrorism* as 'the killing of uniqueness' through acts of violence (2009: 43). War erodes individuality in a multitude of ways. Firstly, the use of conscription and general military operations rely on soldiers relinquishing their individuality for the good of the armed forces and war effort. This is visually reflected in the use of uniforms, for example. Secondly, individuality is lost in the sheer volume of casualties. The atrocities of global violence have slaughtered populations in their millions, and it is incomprehensible to imagine the unique names, lives, and personalities of individuals made invisible in numbers of mass losses. Individuality is lost as bodies become both weapons (soldiers), and targets (casualties). As per Cavarero and noted above, the face is the most unique part of the body, thus Dadaists performed the destruction of bodily and facial integrity to demonstrate the *horrifying* disfigurement of war. *The Gas Heart* by Tzara in 1921 has been interpreted as an example of such theatrical protest (Garner, 2007). This Dada performance, or 'manifestation' as they were called, was played by artists playing the characters of Ear, Mouth, Eye, Eyebrow, Nose, Neck, and Heart, respectively.

Masks were first introduced to Dada by artist Marcel Janco in 1916. Janco referred to his masks as 'Negro,'[19] and these were used to symbolise primitivism alongside drumming to create so-called 'Negro rhythm' (Prevots, 1985; Ball, 1916b). Although Dadaists may have justified this as dismantling the prestige of European culture and art,[20] the conflation of African attributes and dark skin to mysticism or vulgarity is highly problematic. With attributes of the alter-egos used by performers,[21] there were observations that behaviour changed when wearing masks, which were used in improvised dancing (Adamowicz, 2019). Ball described that when wearing Janco's masks, 'something strange happened … bordering on madness' (1916b: 64). Dada collaborator and choreographer, Rudolf Laban, instructed female dancers, described as 'Negresses,' to represent ugliness and deformity (Adamowicz, 2019). A 1916 dance entitled *Negermimus Mime Negre* included music produced by Ball and Janco's masks. It was split into three parts: Flycatching, Nightmare, and Festive Despair. It included shrill music and clumsy, wild, and menacing motifs. Such connotations with the abject rely on hideously racist tropes of African culture (as though such a homogeneity exists) as bizarre and animalistic.

Significantly, the performance of Blackness as primitive not only invites but *requires* the audience to subscribe to these associations (even if temporarily). That is, the horror and decadence of the primitive mask or movement can only induce horror if it already exists in the audience's mind. To reiterate, the performance of Blackness as primitive and bizarre relies on the audience already holding these associations; otherwise, there would be no reaction. Imagine attending an interpretative dance with no context or programme – one would likely find it nonsensical or rely on their own symbolism to construct meaning. A little like the set-up of a comedic punchline, performance draws upon what the audience already knows or thinks. This is not unlike the reaction to Drill balaclavas, informed by existing associations and self-perceptions. Provoking intense and instinctive reactions from audiences enabled Dada to stage these as demonstrative of the attitudes it so scorned (or at least claimed to): 'the artist as the organ of the outlandish threatens and soothes at the same time. The threat produces a defence. But since it turns out to be harmless, the spectator begins to laugh at himself about his fear' (Ball, 1916b: 54). I suggest this is possible only by the contrast of Whiteness, vis-à-vis White performers. This is necessary to remind the audience that it is indeed only a performance. If these dances and phonetic poems were delivered by Black artists, it could arouse anxiety in White audiences already holding problematic beliefs of Blackness. The threat would remain without relief. The audience would not be suspending disbelief through symbolism, instead exposed to a 'reality' they already subscribe to without any contrast to interrupt it. The threat here, is the strange and uncomfortable Blackness, the relief or defence is the White performer who enables separation between performance and reality, revealing its 'harmless' nature, and enabling the spectator to, as Ball puts it, laugh at his fear. Drill, however, which does not easily or purposefully distinguish performance from reality, does not provide relief to counteract racially informed reactions, and is therefore seen as legitimising stereotypes rather than challenging them.

Black Avant-Garde

Although rap music has historically been associated with gangs and criminality (Nelson and Dennis, 2019), UKD has faced increased attempts to target and criminalise the music itself. That is, rather than prosecuting demonstrable acts of violence where injury is directly caused to a victim, Drill itself is classified as a crime, even constituting terrorism (Fatsis 2021; 2022). Social media and engagement with certain rap genres have been enough to register individuals on the police gangs matrix even if no criminal act has been found (Elliott-Cooper, 2021). In the 1990s, Glasgow was considered the hotbed for knife crime in Europe but unlike contemporary discourse, this was not blamed on music nor ethnicity (Williams and Squires, 2021). Today's rhetoric of violence being a Black/rap problem has led to a denial of artistic licence to Black artists whose music is seen as confessional rather than observation. Despite Drillers repeatedly protesting that 'all music is metaphors' (LBC, 2020) and that lyrics should not be taken literally, the intention of individuals – central to criminal prosecution via *mens rea*, is disbelieved or outright ignored. Authorities such as the police are endowed with the authority to reinterpret lyrics as expert witnesses in legal cases. They have a biased investment in finding culpability and making visible their role as preventative and proactive enforcement. Given that the police do not represent the demographic of Drillers as predominantly young, Black people, and are not art historians or music critics, it is unreasonable to provide them with a monopoly of 'translating' the lyrics of an already over-policed population. There are often not enough expert defence witnesses, particularly for such a new genre, which means that the prosecution of rap lyrics may not receive fair and equal representation.

By perceiving UKD only as a threat, opportunities are missed to hear what social issues are being observed and discussed through rap music. Lyrics taken out of context are likely to suggest hostility and violence without insight as to their wider purpose, and whether it genuinely constitutes a serious and imminent threat. The lyrics in *Political Drills (The Media)* by Skengdo, AM and DrillMinister, could be interpreted as threatening politicians and 'bankers' although it would be surprising that the artists would do so on such a high-profile, public platform. Instead, they should be considered as an expression of anger at socio-economic inequality, which the artists hold politicians and financial institutions responsible for. For example, they refer to living off cold baked beans and decrepit estates whilst rent and tax increases and bankers are 'stealing pence' (MixTape Madness, 2019). In his earlier track, *Political Drillin*, DrillMinister (2018) builds his lyrics by quoting politicians such as Jess Phillips MP 'I will knife you in the front,' and former Chancellor George Osbourne 'I will not rest until she's chopped up in bags in my freezer.' The purpose of this song is to highlight the hypocrisy of politicians, condemning Drillers for using violent language whilst using the same speech themselves. 'People were in shock because this was the worst Drill song ever made. It was some of the most violence anyone had ever heard' (LBC, 2020).

This evokes the Dadaist intention to distort and confuse 'as a means of arousing the bourgeoisie to rage, and through rage to a shameface[d] self-awareness' (Richter, 1964: 9). Without being able to see the video's transcription and who is being quoted, the track relies on listeners assuming that the lyrics are the artist's own, only to be shocked when realising their origins and embarrassed for presuming the Driller to be the source of aggression. There is an element of satire and provocation in Drill, similar to Dada, whereby 'the boundaries between stage and audience were similarly violated, as spectators were provoked into reactions that themselves became part of the Dada performance event' (Garner, 2007: 503–504). Dada questioned the complicity of the audience in the violence and apathy around them by provoking as strong a reaction as possible, be this anger, disgust, confusion, or fear. This locates undesirable attributes in the audience themselves, as though this could shock them into protest. This could draw parallels with DrillMinister's 'misleading' listeners to call to account the violence by major power figures, thereby levelling the burden put on Drill. It makes a mockery of the reactionary disdain of Drill if it is not consistently applied to other public contributors.

Aside from subtle or overt political commentary, Drill becomes a tool of resistance simply by existing. To risk the suspended and custodial sentences, restrictive orders, censorship and cancelling, and still make Drill is a defiance of authoritative regulation and control. This includes online fan accounts, which re-upload deleted videos. This is not necessarily thought of as political activism. However, political acts ought not be defined only by their intent, as this places limitations on who can be considered a political actor. Irrespective of personal motivations, something is *being done* when Drillers and their listeners claim a space in the public arena, online or otherwise. Further consideration might be given to the political significance of Drill as an act of refusal. The insistence of Drillers to keep creating, re-uploading, and collaborating reflects a dissatisfaction and rejection of police authority and cultural policing. Similarly, the absence and outlawing of Drill *does something* to regulate public discourse. It silences and ostracises entire communities from public and cultural discourse, whilst justifying policies, which demonise and criminalise an already disproportionately policed group. There is no contradiction in criticising elements of public speech, whilst acknowledging broader perspectives. One can be critical of Drill, without suppressing it entirely.

One of the criticisms levelled against UKD is that it glorifies violence and perpetuates a negative stereotype of Black people for commercial profit. This could refute the avant-garde qualities of Drill by emphasising its profitability. However, nobody exists outside of hegemonic economic infrastructures, and most need to work for money to survive and pay bills. It would be highly problematic to suggest that activism or artistic revolutions can only be led by volunteers without any financial incentive. This would result in public arenas and cultural spaces being represented only by those wealthy enough to create with total financial independence, and this minority is unlikely to reflect those they might claim to empower. Secondly, this financial motivation may be less about making millions, and more about making enough to be stable and have choices beyond the 'road life' Drill speaks of. This is not to say there are not bragging displays or claims of wealth, but even these are premised on it being unexpected, or contrary to the artist's former financial status. It is less common, at least outside of lyrics, for Drillers to claim they are focused on profit at all costs, and more typical that Drillers describe the music as a way out. Digga D explains 'without music, my life would be crazy ... There is nothing else for us to do. Nobody helps us through nothing. We try to help ourselves through music and then they try to take it away from us' (Thapar, 2021). Krept and Konan (2019) unambiguously reflect this in their lyrics describing the story of a teenager entering a cycle of violence due to poverty. After leaving prison he tries to build a music career as a rapper but is constantly censored and obstructed. Consequently, he goes back to the drug economy ('trapping') to survive and is eventually stabbed. The song ends directly and explicitly stating that the responsibility lies with those that banned Drill. This argues a direct causation between censorship and violence. Instead of its slang connotations for drug houses (trap), young people are literally trapped; unable to find legitimate opportunities to make money, financially, and professionally held at the economic periphery.

Conclusion

Dada and Drill share similarities in their favouring of replicability over exclusivity, using violent language (or doing a violent *to* language), and the use of masks. What clearly distinguishes them is the criminalisation of Drill in contrast to the celebration of Dada. I have suggested that this is in part due to the origins of rap music as Black expression, and that racialised artists are obstructed by social assumptions that conflate Blackness with actual, rather than symbolic, violence. Although rap music has always received uneven scrutiny and fear, UKD is restricted in such a way that the music itself increasingly constitutes a crime. This simultaneously means that, irrespective of the artist's political intent, engagement with the genre could constitute, or be perceived to constitute, resistance. Drillers and their audiences have become the new 'suspect communities' to be monitored and controlled. This is unlike the policing of other art forms, which, like Dada, enjoy the freedom to deliberately rouse anger and aggression in its audiences. The shunning of Drill from public space suppresses the chosen expression of already disenfranchised groups and silences their talking points, such as experiences of police, racism, youth culture, and city life. Although Drill does not have a clear and consistent campaign like Dada's anti-war message, it could be said to be reflecting the reality of wars on gangs, too often translating to wars on urban, Black communities. There may be anti-police sentiment as there is across rap music. Ironically this is consolidated and legitimised through the endless policing of arts and culture. Artistic licence to embellish, to provoke, is denied to young Black artists, whose aggression cannot be imagined by the public as symbolic because they already associate them with violence. Therefore, considering alternative interpretations of Drill is instrumental in challenging the disproportionate policing of arts and culture created by racialised communities. By comparing UKD to Dada, I am suggesting it can be reimagined as the new avant-garde, which, without denying its aggressive visual-linguistic content, is a reflection of violence, not the cause. Condemning it so severely has implications for access to public space and causes social injury by ostracising contributors who may otherwise disengage. It also suggests that current policies, which intend claim to prevent street violence, are ineffective so long as they target music instead of the socio-political circumstances within which it thrives.

Notes

1 I use UK Drill and UKD interchangeably throughout.
2 Artists who make Drill.
3 Balaclava.
4 Notable figures who fled to Switzerland include Lenin, who coincidentally lived opposite *Cabaret Voltaire* but did not participate (Garner, 2007).
5 This could be compared to hip-hop, which by its definition includes music, dance, fashion, and graffiti.
6 As discussed in the introduction of this book, it is important not to dismiss the political significance of movements and practices, which are not driven by specific ideologies or activist intent.
7 This creates rhythm from the last word of each line, whereas ABAB for example, aligns every other line.
8 Early 90s hip-hop also faced emphasis on White listenership. See Krims (2000).
9 The name of Dr. Dre and Ice Cube's 1994 song, which coincidentally includes the lyrics 'they call me dada.'
10 Incriminating oneself by confessing to crime in songs.

11 However, the visuals for this video are not clear cut given that the artists stand in front of signs reading 'Officer I've Done Nothing,' and hold medical blood bags, which suggests they may be assisting the policeman or even transfusing their own blood. Given the title of the song *Political Drills (The Media)* and the collaboration with DrillMinister, this track may be making a point about assumptions in the media that Drill is violent, by suggesting they are helping rather than harming the policeman. The lyrics include references to the 'Momo Challenge,' which was a social media hoax inducing mass panic following rumours that a user called 'Momo' was inciting children to harm themselves or others, even though this was found to be untrue. See Waterson (2019).
12 Lyric in *Political Drillin* (DrillMinister, 2018).
13 This has since between scrapped. See BBC (2017) 'Form 696: "Racist police form" to be scrapped in London.'
14 This was rebranded in 2013 to consider all shootings in London. See Young et al. (2014).
15 See Amnesty International (2018). As this article was being written, news broke in November 2022 that the Metropolitan Police have been forced to revise the Gang Matrix following a High Court challenge instigated by Awate Suleiman, who, I note, is a rapper.
16 Whereby an individual is jointly charged as encouraging or assisting crime perpetrated by someone else.
17 See Home Office (2017).
18 This re-enforces my earlier example of Dada insulting its audience.
19 I was initially reluctant to cite this language given its offensive connotations but decided to do so to confront their use by Dadaists and so as not to evade holding artists accountable.
20 Examples include Hannah Höch who produced a series of photomontages blurring African masks with European women. This was problematic as removing the masks from their ethnographic context was used to symbolise disorganisation and horror, somewhat like a freak show (Adamowicz, 2019). However, this may have been precisely her intention; to challenge the normative femininity of the European body and the deviancy of the African figure.
21 I am cautious here not to directly equate it to the alter-egos of Drillers, as there are different power relations at play when White performers use alter-egos to perform their interpretations of Blackness, in this case symbolising primitivism.

References

Adamowicz, E. (2019) *Dada bodies: Between battlefield and fairground*. Manchester: Manchester University Press.

Aitkenhead, D. (2014) Carole Duggan interview: 'I'm not going away and they're not shutting me up'. *Guardian*, 28 Feb, available at: www.theguardian.com/uk-news/2014/feb/28/carole-duggan-interview-not-going-away. Accessed 7 Nov 2022.

All-Party Parliamentary Group (APPG) (2019) *There is no protection on the streets, none. Young people's perspectives on knife crime*. www.barnardos.org.uk/sites/default/files/uploads/APPG%20on%20Knife%20crime%20-%20Young%20people%27s%20perspective%20August%202019.pdf. Accessed 22 Dec 2019.

Amnesty International (2018) *Secrecy, stigma, and bias in the Met's Gangs Database.*
Arnason, H. H. (1969) *A history of modern art: Painting, sculpture, architecture.* London: Thames and Hudson.
Attali, J. (1985) *Noise: The political economy of music.* Minneapolis: The University of Minnesota Press.
Aurbacher, H., Maier, T., Liska, E. (1916) [2015] Hugo Ball: Gadji Beri Bimba. *Mr. DSCH.* www.youtube.com/watch?v=aiKHSeDlU1U&ab_channel=Mr.DSCH. Accessed 1 Aug 2022.
Ball, H. (1916a) *Dada manifesto.* In H. Ball. J. Elderfield (ed) *Flight Out of Time: A Dada Diary.* J. Elderfield (ed). California and London: University of California Press, pp. 219–221.
Ball, H. (1916b) *Romanticism: The word and the image.* In H. Ball. J. Elderfield (ed) (1996) *Flight Out of Time: A Dada Diary.* J. Elderfield (ed). California and London: University of California Press, pp. 50–132.
Ball, H. (1917a) (2020) *Hugo ball – Karawane. Shony.* www.youtube.com/watch?v=PWKP5OAsYZk&ab_channel=Shony. Accessed 1 Aug 2022.
Ball, H. (1917b) [1996] *Kandinsky.* In H. Ball. J. Elderfield (ed) *Flight Out of Time: A Dada Diary.* J. Elderfield (ed). California and London: University of California Press, pp. 222–234.
Baraka, A. [1965] (2009) Amiri Baraka reads black art. *Hoodoojazz.* www.youtube.com/watch?v=Dh2P-tlEH_w&ab_channel=hoodoojazz. Accessed 20 Aug 2022.
BBC. (2011) Public enemy – Prophets of rage documentary. *Arm the Creative.* Available at: www.youtube.com/watch?v=1fM_VXPZqBg&ab_channel=ArmTheCreative2. Accessed 25 July 2022.
BBC. (2017) Form 696: 'Racist police form' to be scrapped in London. *BBC News*, 10 Nov 2017, available at: www.bbc.co.uk/news/uk-41946915. Accessed 11 Dec 2024.
BBC. (2018) Camberwell stabbing: Drill rapper Incognito killed. *BBC*, 2 Aug, available at: www.bbc.com/news/uk-england-london-45039590. Accessed 20 Aug 2022.
BBC. (2021) Birmingham Brothers banned from drill music videos glorifying crime. *BBC*, 27 May, available at: www.bbc.co.uk/news/uk-england-birmingham-57268903. Accessed 2 Feb 2022.
Benjamin, W. (1921) [2021] Toward the critique of violence. In P. Fenves and J. Ng (eds) *Toward the Critique of Violence: A Critical Edition.* Stanford: Stanford University Press, pp. 39–64.
Brighton and Hove Community Safety Partnership. (2020) *Community Safety and Crime Reduction Strategy 2020–23.* Available at: www.safeinthecity.info/sites/safeinthecity.info/files/sitc/Brighton%20%26%20Hove%20Community%20Safety%20and%20Crime%20Reduction%20Strategy%202020-2023.pdf. Accessed 1 April 2022.
Buckley, J. and Tsai, J. (2020) Dada futures: Introduction. *Dada/Surrealism*, 23: 1–7.
Canton, W. (2022) I Spit Therefore I Am: Rap as Knowledge. *Interfere*, 3: 58–81.
Cavarero, A. (2009) *Horrorism: Naming contemporary violence.* New York: Colombia University Press.
Chang, J. (2007) *Can't Stop Won't Stop: A History of the Hip-Hop Generation.* London: Ebury Press.

Dancemyday TikTok. (2022) *You don't look British | If you don't know me I'm M to the B | Chav make up compilation.* Available at: www.youtube.com/watch?v=Iuox i17LGK4&ab_channel=DancemydayTikTok. Accessed 23 Oct 2022.

Dodd, V. (2019) Mark Duggan shooting report challenged by human rights groups. *Guardian*, 5 Dec, available at: www.theguardian.com/uk-news/2019/dec/05/mark-duggan-shooting-report-challenged-by-human-rights-groups. Accessed 7 November 2022.

DrillMinister. (2018) *Political drillin.* Available at: www.youtube.com/watch?v=spJo RLpDLLM&ab_channel=LinkUpTV. Accessed 20 Jan 2022.

Duchamp, M. (1966) [2008] Duchamp interviews. Available at: www.youtube.com/watch?v=7CFQY0Yf1iI&ab_channel=bamchum. Accessed 20 Oct 2022.

Elliott-Cooper, A. (2021) *Black resistance to British Policing.* Manchester: Manchester University Press.

English Frank. (2011) Warm up sessions [S2.EP13]: SBTV. Available at: www.youtube.com/watch?v=ZhkoegM9HQE&t=88s. Accessed 27 April 2020.

Falkner, S. (2021) Knife Crime in the Capital: How gangs are drawing another generation into a life of violent crime. *Policy Exchange*, 11 Oct, available at: https://policyexchange.org.uk/publication/knife-crime-in-the-capital/. Accessed 28 July 2022.

Fatsis, L. (2019) Policing the beats: The criminalisation of UK drill and grime music by the London Metropolitan Police. *Sociological Review*, 67: 1300–1316.

Fatsis, L. (2021) Sounds dangerous. In N. Peršak and A. Di Ronco, *Harm and Disorder in the Urban Space: Social Control, Sense and Sensibility.* London and New York: Routledge, pp. 30–51.

Fatsis, L. (2022) Arresting sounds: What UK soundsystem culture teaches us about police racism and public life. In Charles, M. and Gani, M.W. (eds) *Black Music in Britain in the 21st Century.* Liverpool: Liverpool University Press, pp. 181–198.

Frith, S. (1981) 'The magic that can set you free': The ideology of folk and the myth of the rock community. *Popular Music*, 1: 159–168.

Garner, S. (2007) *The Gas Heart*: Disfigurement and the dada body. *Modern Drama*, 50: 500–516.

Griffiths, B (2021). Knife Crime on Rise Stabbings soar as kids turn to knife crime and gangs during pandemic. *Sun*, 13 Feb, available at: www.thesun.co.uk/news/14043 382/kids-turn-to-knife-crime-gangs-pandemic/. Accessed 2 April 2022.

Grindon, G. (2011) Surrealism, dada, and the refusal of work: Autonomy, activism, and social participation in the radical avant-garde. *Oxford Art Journal*, 34: 79–96.

Haakenson, T. (2019) 1968, Now and Then: Black Lives, Black Bodies. *Cultural Critique*, 103: 75–83.

Harper, T. (2017) 11 stabbed every day as knife crime soars. *Times*, 19 Feb, available at: www.thetimes.co.uk/article/11-stabbed-every-day-as-knife-crime-soars-r5726p 8hk. Accessed 12 Sep 2022.

Home Office. (2017) Police powers and procedures, England and Wales, year ending 31 March 2017.

Huncho, M. (2020) M Huncho Interview: Unmasked Thoughts | The Perspective @ AmaruDonTV. *Amarudontv*. Available at: www.youtube.com/watch?v=avt7pj98 yD4&t=84s. Accessed 4 Oct 2022.

IOPC: Independent Office for Police Conduct. (2022) Statement read out at opening of inquest into the death of Chris Kaba. Available at: www.policeconduct.gov.uk/news/statement-read-out-opening-inquest-death-chris-kaba. Accessed 26 Oct 2022.

Jones, D.W. (2014) *Dada 1916 in theory: Practices of critical resistance.* Liverpool: Liverpool University Press.

Keith, J. (2020) The evolving sound of UK Drill. *DJ Mag*, 21 Oct, available at: https://djmag.com/longreads/evolving-sound-uk-drill. Accessed 19 Sep 2022.

Keyes, C. (1996) At the crossroads: Rap music and its African Nexus. *Ethnomusicology*, 40: 223–248.

King, J. (2019) Rap any drill song word for word to win £100 (Roadman Edition). Available at: www.youtube.com/watch?v=BTbURHOXy-4&ab_channel=JaydenKing%28Canking%29. Accessed 23 Oct 2022.

Krims, A. (2000) *Rap music and the poetics of identity.* Cambridge: Cambridge University Press.

Krept and Konan. (2019) *Ban Drill.* Available at: www.youtube.com/watch?v=1rZWn_vcGac. Accessed 28 Oct 2022.

LBC. (2020) *Drillminister on why he's running for Mayor: "I'm the realist Londoner in this election".* Available at: www.youtube.com/watch?v=Iu-d1QkJns8&ab_channel=LBC. Accessed 11 Dec 2020.

Lee, M. (2022) This is not a drill: Towards a sonic and sensorial musicriminology. *Crime, Media, Culture*, 18: 446–465.

Lynes, A., Kelly, C. and Kelly, E. (2020) Thug Life: Drill music as a periscope into urban violence in the consumer age. *British Journal of Criminology*, 60: 1201–1219.

MixTape Madness. (2019) Skengdo x AM x Drillminister – Political Drills (The Media) (Music Video) | @MixtapeMadness. Available at: www.youtube.com/watch?v=qtlziOxKjcY. Accessed 8 Nov 2022.

Moore, S. (2019) Drill group Zone 2's video pulled by YouTube after allegedly naming murder victims. *NME*, 10 Dec 2019, available at: www.nme.com/news/music/drill-group-zone-2-no-censor-video-pulled-by-youtube-2585817. Accessed 11 Dec 2024.

Murphey, T. (1998) Motivating with near peer role models. In B. Visgatis (ed). *On JALT'97: Trends & Transitions*, Tokyo: The Japan Association for Language Teaching, pp. 201–206.

Nelson, E. and Dennis, A. (2019) *Rap on trial: Race, lyrics and guilt in America.* New York/London: The New Press.

Noisey. (2022). LD: The Godfather of UK Drill Returns | Noisey Raps. www.youtube.com/watch?v=PIQW9_n1TmI. Accessed 10 Dec 2022.

Office for National Statistics (ONS). (2020) *Homicide in England and Wales: year ending March 2019.* Available at: www.ons.gov.uk/peoplepopulationandcommunity/crimeandjustice/articles/homicideinenglandandwales/yearendingmarch2019#which-groups-of-people-were-most-likely-to-be-victims-of-homicide. Accessed 2 Dec 2020.

Prevots, N. (1985) *Zurich dada and dance: Formative ferment. Dance Research Journal*, 17: 3–8.

R6. (2016). *redruM reverse (Prod. By Carns Hill).* Available at: www.youtube.com/watch?v=QVQFilmqa7k. Accessed 20 Nov 2019.

Rahim, Z. (2019) Men who continued to make drill music 'inciting and encouraging violence' despite injunction are sentenced. *Independent*, 18 Jan, available at: www.independent.co.uk/news/uk/home-news/drill-music-crackdown-london-met-police-lambeth-gangs-knife-crime-a8735046.html. Accessed 20 Nov 2019.

Richter, H. (1964) [1997] *Dada art and anti-art*. London: Thames & Hudson Ltd.

Rose, T. (1994) *Black noise: Rap music and Black Culture in contemporary America*. Hanover and London: University Press of New England.

Stuart, F. (2020) *Ballad of the Bullet: Gangs, drill music, and the power of online infamy*. Princeton and Oxford: Princeton University Press.

Thapar, C. (2019) Bally on me: Why UK rappers cover their faces. *The Face*, 7 May, available at: https://theface.com/music/bally-on-me-why-uk-rappers-cover-their-faces. Accessed 4 Sep 2022.

Thapar, C. (2021) Free Digga D! A rare interview with the most influential British rapper of our time. *GQ Magazine*, 11 Oct, available at: www.gq-magazine.co.uk/culture/article/digga-d-interview. Accessed 12 Dec 2022.

Trend Centrl. (2020) *The Zeze Millz Show: Ft Loski – "These Guys Doing Drill Aren't Really Gangster's"*. Available at: www.youtube.com/watch?v=78SnPpSBXMo&ab_channel=TRENDCENTRL. Accessed 1 Jan 2022.

Tzara, T. (1918) *Dada manifesto*. Available at: https://writing.upenn.edu/library/Tzara_Dada-Manifesto_1918.pdf. Accessed 20 Aug 2022.

UK Parliament. (2018) Organised crime: Young people's safety. *Hansard*, 5 Sep 2018, available at: https://hansard.parliament.uk/commons/2018-09-05/debates/18090555000001/OrganisedCrimeYoungPeople%E2%80%99SSafety. Accessed 11 Dec 2024.

Vozick-Levinson, S. (2022) New York city mayor Eric Adams declares war on drill rap. *Rolling Stone*, 11 Feb, available at: www.rollingstone.com/music/music-news/mayor-eric-adams-drill-rap-1299108/. Accessed 7 April 2022.

Waterson, J. (2019) Viral "Momo challenge" is a malicious hoax, say charities. *Guardian*. www.theguardian.com/technology/2019/feb/28/viral-momo-challenge-is-a-malicious-hoax-say-charities

Williams, E. and Squires, P. (2021) *Rethinking Knife Crime Policing, Violence and Moral Panic?* Switzerland: Palgrave Macmillan.

Young, T., W. Fitzgibbon and Silverstone, D. (2014) A question of family? Youth and gangs. Youth Justice, 14: 171–185.

4
MEMETIC FEMINISM AND TIKTOK

Kathryn Zacharek and Wanda Canton

Although the social media app, TikTok, might not typically be imagined as a feminist space, emerging memes are uniting young women online. Studies of social media as 'digital dissidence' have included discussions of the use of humour to laugh at (predominantly governmental) power (Matsilele and Mututwa, 2021). TikTok, however, as a relatively new platform and primarily used by a much younger demographic than other platforms, has received less attention. Further, whilst channels such as YouTube, Twitter, or Facebook have elevated the voices of political activists, this chapter is concerned with young women and girls who do not necessarily identify themselves as activists; nevertheless, we aim to demonstrate how they play a role in countering online misogynistic discourse. As *Sonic Rebellions* considers how sound can be used to advocate for justice, it also acknowledges how sound is politicised and regulated, through hegemonising sonic architectures. In this case, we explore how speech, memes, and podcasting are used both to consolidate and resist gender inequalities. In particular, we identify parody podcasts, which mock the popular misogynist 'alpha' podcast, *Fresh & Fit*. Although this podcast is not an isolated case of rampant sexism online, it serves as a commercially successful example through which we can locate contemporary gender politics and memetic responses. Misogynist and alpha are used interchangeably throughout; misogyny being distinct from general discriminatory beliefs (sexism), as an active and coercive attempt to silence women, primarily perpetrated by individuals (Manne, 2018). The disciplinary practice of misogyny is constructed through discourse, and its regulatory function. By mimicking the misogynist language of these 'alpha' podcasts

DOI: 10.4324/9781003361046-5

and inducing laughter in the audience, parodic TikTok memes counteract the legitimacy and oppressive discourse of the manosphere.[1]

We utilise discourse analysis of non-literary text (podcasts and memes) to consider how they operate as power/resistance in Foucauldian terms and through a feminist lens. We begin by contextualising misogynist podcasts, and their concerning attitudes towards women. As of August 2023, *Fresh & Fit* have been removed from the YouTube Partnership Program and their videos have been demonetised (Dodgson, 2023), demonstrating the relevance of this discussion and ongoing debates at the time of writing this chapter. Self-defined alpha men (as opposed to the subordinate 'beta' men) and podcasters favour machoistic approaches of interrupting women (or as they prefer, 'female') guests, overexplaining (mansplaining) and simplifying arguments to the point of incomprehensibility, using bio-evolutionary discourse to claim sexual difference, as though contemporary society could still be premised on the preferences of prehistoric cavemen. These ideologies cross over into real-world relationships and have serious ramifications as will be discussed. These include, but are not limited to, allegations of serious assault, human trafficking, multi-level marketing (MLM) schemes, and growing concerns about the influence these podcasts have on young men. Drawing upon Foucauldian concepts of power and resistance, we explore how these podcasts contribute to patriarchal consolidations and discourse. We refer to power 'cells' as discursive networks, which Foucault explains manifest in multiple configurations. We utilise the term 'cell' meaning network or organism, to draw parallels with our discussion of memetic/ genetic viral attributes. It also subtly alludes to *incel* (involuntary celibate); an online community of men who resent women for not fulfilling their sexual expectations.

However, Foucault reminds us that wherever power is found, so too is resistance. We therefore turn our attention to the use of TikTok counter-memes, focusing primarily on the use of parodying content. We argue that its comedic attributes, which we differentiate from social media 'trends,' enable the memes to build community amongst women online and, by evoking laughter in audiences and creators, offers opposition to misogynist discourse. Key signifiers of rationality and science are used in an attempt to garner authority, but are contradicted by the belly-laughs of women who refuse to concede. References throughout will primarily focus on English-speaking content from the US as the memes generally respond to American podcasts. However, given that social media content transcends national borders, and there are international concerns about public figures such as Andrew Tate, himself British–American, we will also discuss some of the UK responses. We refer to a number of internet sources, including TikTok posts, and mainstream media, including newspapers, given our attention to popular media discourse and its impact on gender politics.

Men and Their Microphones

Fresh & Fit was founded in 2020 by Myron Gaines and Walter Weekes, an ex-Homeland Security agent-turned-investor and record producer, respectively. Their Spotify description claims their podcast tells the 'TRUTH' (original emphasis) in dating, fitness, and using an 'evidence-based approach' to 'fully optimise the dating market.' The use of pseudo-science, economic, and business discourse[2] attempts to garner authority for the podcast and cloaks their misogyny in terms that suggest objective legitimacy. They claim to examine 'female psychology' as though women, referenced only by their sex, are victims to subordinate bio-evolutionary impulses. The *Fresh & Fit* hosts infantilise women through their choice of language, often referring to their guests as 'girls,' with Gaines and Weekes depicting women as immature and irresponsible. Throughout the *Fresh & Fit* podcast, women are positioned as the property of men, owing their loyalty to the 'high value' men who they claim sustain women by way of monetary and physical protection. The podcast defines the value of men by their finances, and the value of women by their chastity or selective modesty. For example, Gaines and Weekes argue women in relationships shouldn't have social media accounts such as Instagram, as they claim it primarily serves to exhibit her attractiveness to other men (Reid, 2022). A woman's body, according to *Fresh & Fit*, should only be for her male partner to see. This is framed in the discourse of business ('risk') and family values, despite it being a clear attempt to justify the control of women:

> If you really ask guys what they want, they want a girl that's loyal, submissive, and not a hoe, but not being a hoe weighs a lot, and because we take an enormous risk when we take a girl on that has a promiscuous past and men don't want to be embarrassed.
> *(Fresh & Fit, 2021)*

Such claims are based on 'a bunch of studies out there ... a bunch of stats,' used to make statements such as 'the more promiscuous a woman is, the less likely she's gonna be a good *wife* and more importantly a good mother to your child' (ibid; emphasis added). Although sources are not usually provided for what the hosts present as facts and information, Gaines and Weekes do notably cite one online, non-scholarly source entitled the 'Walkaway Wife Syndrome' (Divorce.com, 2023) to support their arguments about promiscuity and divorce. However, Gaines and Weekes completely misconstrue the article, which makes no reference to the number of sexual partners a woman has had as an indicator of divorce. Rather, the article argues that communication breakdown, emotional disconnection, and disinterest on the part of both parties are the primary causes of a marriage ending.

Alpha podcasters demonstrate a clear and consistent fixation on women's sexual lives, obsessing over their 'body count' in reference to their number of sexual partners. 'If she has a promiscuous past, she might be a fantastic person, she might be smart, she might be intelligent, whatever. The most important foundational root though is tainted' (Fresh & Fit, 2021). In one extreme example, they conflate sexual violence to raising another person's child. Indicators of possession are italicised below.

> Yeah, so if it's that pain that you feel when you're getting taken advantage of, and someone is doing things to you that you don't want, [this is] how men feel about raising a child that they find out *isn't theirs*, okay? We get raped another way. That's why men don't want to deal with girls that are hoes because if you deal with a girl that's promiscuous and has a past and *you have* a child with this woman, there's a good chance that child might not *be yours*, and that is devastating to a man. So that is why a woman's past matters so much to men, it's just that we live in a society nowadays that shames men for holding women to standards.
>
> *(Fresh & Fit, 2021; emphasis added)*

There are growing concerns about how online behaviour seeps into the everyday or IRL.[3] In 2023, *Fresh & Fit* were gaining 10,000 new subscribers a week (Playboard, 2023). Whilst it is not the case that every subscriber will agree with the podcasters' views, it certainly provides them with a large enough platform to reach those who do. Rather than being constrained to the fringes of the internet, the alpha podcast is creeping into the mainstream.

British–American Andrew Tate, a fellow podcaster and frequent guest on *Fresh & Fit*, is notoriously misogynistic both online and off. His podcast, *Tate Speech*, is now only available in the format of short clips on one website[4] and has amassed 1.49 million followers. Before Andrew Tate was de-platformed from Instagram, he had 4.7 million followers. After a five-year ban on Twitter, once Tate was reinstated onto the platform by Elon Musk, he gained a million followers in 24 hours (Das, 2022a; Shrivastava, 2022). In 2022–2023 Tate's videos were watched 13 million times by a global audience, attesting to the volume and reach of his influence (Oppenheim, 2023c). He has used his online following to promote what his supporters refer to as motivational courses, such as 'Hustler University,' which has been criticised for not only being a pyramid scheme but encouraging men to create coercive web-cam businesses by lying to and manipulating women. This, in some cases could amount to, and has been accused of, human trafficking (Das, 2022a; Dahir, 2023). Like *Fresh & Fit,* Tate's narratives define women as property, and there is well-documented evidence of how he has used his financial position to control and exploit women: 'if I have a responsibility over her, then I must have a degree of authority' (Tate cited in Mustafa, 2022).

Tate's history of direct violence against women first came to public attention in 2016, when he was expelled from the UK reality TV show, *Big Brother*, after one video was published showing him beating a woman with a belt, and another in which he demands a woman count the bruises he had caused her (Das, 2022b). In 2022 he was held on accusations of human trafficking in Romania (Wright & Murphy, 2022), and in 2023 three women began pursuing legal proceedings against him, alleging that between 2013 and 2016 they were victims of sexual and physical assaults (Oppenheim, 2023a; Thomas, 2023). Concern about Tate's influence on young men continues to escalate with some schools in the UK delivering lessons, or 'Andrew Tate assemblies,' to school children, and advising parents on how to discuss his problematic views at home (Mathers, 2023; Evans, 2023). A 2023 poll found that more teenagers aged 16–17 were able to identify who Andrew Tate is than the current British Prime Minister (Oppenheim, 2023b). Although it is not necessarily surprising that younger people take more of an interest in popular culture than Politics with a capital P, it does highlight the significance of what they are consuming online. Tate has since been banned or had his content removed across YouTube, TikTok, Facebook, and Instagram for violating community policies regarding 'dangerous individuals' (Paul, 2022). Given the notoriety that Tate brings to these platforms, such decisions to remove him suggests the concerns far outweigh the benefits of his online presence, such as profit and revenue to the host app, who gain financially from controversial users.

Sound then, has been used via podcasting to platform oppressive attitudes towards women. Whilst misogyny is nothing new, in an era of globally accessible social media, podcasts can quickly reach mass and instantaneous audiences. Podcasts can transcend the barriers of borders, finance (given that they are generally free to consume) and managing agencies or gatekeepers, as platforms including YouTube and iTunes enable user-generated content, which do not require high levels of specialism. It should be noted that this does not mean that podcast production is a free-for-all. Podcasters are still subject to the host platform's community guidelines, which are enforced through suspensions and de-platforming. Sound is further used as a mechanism of power through the production and style of the podcasts themselves. The podcasters of *Fresh & Fit* use speech as sound to silence their women guests, aggressively demanding they leave when they challenge the hosts, and frequently use a soundboard to interrupt them with air horns, mocking music, or outright silencing their voices with overlaying audio clips. It is notable how aggressively Andrew Tate speaks, over-emphasising if not shouting, using derogatory slurs for women and on occasion brandishing weapons, with machoistic body language including gestures of hitting and strangling women (Das, 2022b).

Power Cells

Power, according to Foucault (1976), is not static but diffused through multiple arms or infrastructural organs; meaning there are several 'cells' (our term) to any given body of power, including discursive networks of speech and imagery. There is no single centre or head of power, rather cells that carry out a variety of functions to consolidate its overarching and expansive power. This draws similarity to the concept of the *body politic*, utilised by political philosophers dating back to Plato, as an analogy for the constitution of the ideal state. For Plato, the metaphor of the body is used to justify his hierarchal and tripartite ideal system in the *Republic*; philosopher kings (or rulers) are the rational head, the warrior class corresponds to the chest, and the workers as the stomach (Plato and Lee, 2007). Thomas Hobbes' *Leviathan* (1651; Hobbes and Macpherson, 1981) maintains that sovereign power is absolute and that the head of the state should not have less power than that of the rest of the body, as it is the role of the sovereign to maintain order and prevent society from descending into a state of all-out war. However, both these representations of the body politic conceive of power as a monolithic entity, whereas power for Foucault is dispersed. Parallels can be drawn between different body parts/institutions carrying out specific functions, but power in Foucauldian terms, rather than being a top-down operation, functions more like blood cells circulating throughout the body as a complex network of veins. Networks of cells are strengthened by being repeated, and expand as they interact with other discursive practices. Patriarchy does not operate at the level of a single institution. Rather it is a collation of ideological and discursive nodal points,[5] which are consolidated by the repetitive reinforcement of cellular text (the components of a network, such as language, dress, family). This, in turn, creates hegemonising norms and standards (hence the emphasis on tradition and biology), which in a patriarchal context includes oppressive gender roles. These are regulated through discursive associations and speech patterns, which re-enforce associations of masculinity and femininity. Social media can be understood as a cellular component of contemporary patriarchal power, reproduced and regulated by misogynistic podcasters. Podcasters regulate, reproduce, and discipline discursive patterns on their platforms. This aims to legitimise or normalise their ideology as natural, factual knowledge. Cells of power dictate the terms of knowledge production regarding what is claimed to be true, and the ways truth can be known. The influence that podcasters wield encourages (some) young men to repeat or emulate misogynistic discourse, thus stabilising the signification of masculinity as wealth, fitness, aggression, and sexual prowess.[6]

Andrew Tate is one (of many) creators who have become successful by claiming to be an expert in masculinity and teaching men how to become the ideal 'alpha male.' Tate places limitations on who constitutes an alpha male, and gives the impression that masculinity is something which can be learned and controlled. One of his key metrics for becoming an alpha is financial success, and this can be supposedly achieved by enrolling on courses in his Hustler's University. This positions Tate, and his followers, as experts or producers of knowledge in the context of gender, and actively seeks the conformity and regulation of other men (including depreciating critical men as subordinate betas). This in turn consolidates hegemonic masculinity,[7] which, by definition, subordinates some men and all women.

Tate's exploitation of the TikTok algorithm has enabled him to perpetuate his brand of misogyny. His Hustler's University provides financial incentives for participants to recruit others into its programmes. This includes training to become an 'affiliate marketer' by editing short clips of Tate and posting them across social media to drive traffic and generate views (Hunter, 2022). Essentially, an army of copycat accounts[8] were produced to share his content, and the individuals running them were financially (and perhaps ideologically), motivated to do so. In turn, this produced more profit for Tate as he was able to reach more people, with new candidates willing to pay the £39 monthly subscription to take part in his MLM courses (Das, 2022b). Akin to the attention economy highlighted in the preceding chapters of this book (see Kaur; Canton), an 'outrage economy' (Shrimsley, 2015) has developed whereby controversial content produces engagement with social media posts, thereby bolstering it on the algorithm's recommended watch lists. The more outrageous Tate's content became, the more reaction he garnered, and the more it was shared and promoted. Even opponents of Tate are made somewhat inadvertently complicit by sharing his material, as their otherwise negative reaction nevertheless generates benefits for him. Tate demonstrates a clear and conscious effort to make his voice loudest and his wallet bigger. As mentioned, his controversy is profitable as a big following or interest can lead to paid brand sponsorships and advertising deals. In another example, in *Fresh & Fit*'s Spotify show notes, it states that vitamin and protein supplement company, Gorilla Mind, and the erectile dysfunction medication supplier, Blue Chew, have sponsored past episodes.

However, Foucault (1976) explains that wherever there is power, there is resistance. Or as per our later discussion; where there are memes, there are counter-memes. Women on TikTok actively seek to subvert the algorithmic calculations, by using the same hashtags as alpha podcasters, thereby flooding the usually misogynist #alpha, #freshandfit or indistinct but popular #podcast feeds with their counter-ridicule.⁹ This has the potential to intercept users searching for Gaines, Weekes, and Tate, offering alternative perspectives to weaken the potency and dominance of their misogyny, hopefully dissuading prospective followers. At the very least, it serves to obstruct the normalisation of the rhetoric by showing dissent. Some reports have suggested that the feminist counter-memes have raised awareness of podcasts such as *Fresh & Fit* to the extent that a surge in complaints against the channel for hate speech resulted in a suspension from YouTube (Glaze, 2022). It is not the purpose here to advocate for a stringent policing of online space, rather to highlight how the use of humorous memes can lead to additional actions to level the online speaking field. It may also serve to expand the membership of who can be considered feminist or politically active, as TikTok provides an alternative way to engage with socio-political and cultural issues for those that might not otherwise contribute. Returning to Foucault (1976), resistance is always 'inside' power. Given power is versatile, expansive, even creative, so too is the opposition. As power is not a monolithic entity, resistance can take on a variety of forms that challenges existing different power cells. Therefore, women mocking alpha podcasters online is a resistance on their turf, as it were. That is, online misogyny is also opposed online. By balancing the volume and pervasiveness of alpha male content, the memetic response produces a counter-discourse.

Källstig and Death (2021) highlight the political agency of comedy by disrupting the reproduction of power discourses and offering counter-discourse instead, particularly when used by those who are targeted by the oppressive material. Although they focus on stand-up, it is nonetheless transferable to memetic humour. They describe 'ambivalent mockery' as staging both the tragedy and comedy of oppressive power relations. Crucially, they note comics' ability to transcend the expected behaviours or actions of activists:

> It is neither the 'pure resistance' of the revolutionary subject, nor simply the co-opted reproduction of neoliberal hegemony, but an ambiguous and multifaceted performance which reveals important contradictions in contemporary forms of power and resistance.
> *(Källstig and Death, 2021: 340)*

Although women TikTokers do not necessarily explicitly identify as feminists, 'feminist activism may not need to be announced as explicitly *feminist* in order to function as such' (Keller, 2012: 441; original emphasis). Arguably, the outcome of better representation for and of women online, equality of speech/sound, and dismantling the hegemony of misogynist discourse can have feminist repercussions irrespective of whether it is intended as such. Throughout this book, the authors discuss a range of practices, which are not always conceived as outright political activism, nevertheless demonstrating their political significance and potential. This may also be true of TikTok, wherein the creation and distribution of feminist memes is a practice of 'culture jamming,' which disrupts mainstream political and cultural narratives at a grassroots level (Rentschler and Thrift, 2015). Culture jamming involves the planned or spontaneous creation of images/memes/practices that encourage viewers to question how media shapes societal norms and values.

The Speaking Meme

Despite attempts to bully and humiliate their guests or critics that challenge the blatant misogyny of alpha podcasters, women are taking to TikTok to 'clap back'.[10] The app, which originally started as a space for user-uploaded dance videos, has become a central platform for distributing information, campaigning, and as a go-to place for crowd-sourced social education. TikTok has become the most popular news source for 12–15-year-olds in the UK (Farah, 2023). Sixty percent of the app's users are Generation Z (Stahl and Literat, 2022); those born after the mid-90s who have grown up with accessible internet and portable digital technologies. Fifty-five percent of TikTok content creators are women (Ceci, 2022), demonstrating the active participation of women and girls on the platform.

The terminology of 'memes' first appeared in biologist Richard Dawkins' book, *The Selfish Gene* (1976).[11] He coined the term as a cultural analogy to genes. That is, the meme transmits *cultural* information as opposed to *genetic*. Like its biological counterpart, the meme mutates and evolves in the process of repetition (Denisova, 2019). In parallel to genetic inheritance, cultural ideas and practices are passed from one generation to the next. In the early 90s, Mike Godwin referred to the 'internet meme' in an article for *Wired*, describing the use of memes on message boards and again comparing it to genetic transmission:

> A 'meme,' of course, is an idea that functions in a mind the same way a gene or virus functions in the body. And an infectious idea (call it a 'viral meme') may leap from mind to mind, much as viruses leap from body to body.
>
> *(Godwin, 1994)*

A meme is similar to a 'trend,' as both replicate a template format, song, or theme, but memes tend towards intentional humour, alongside the individual user's personal adaptation of an original image or text (Rentschler and Thrift, 2015). Memes do not inherently belong to any political or cultural affiliation. Instead, they are conduits filled with meanings shaped by and embedded in the culture where they are used (Denisova, 2019). Significantly, Godwin

(1994) referred to 'memetic engineering' from the outset, 'crafting good memes to drive out the bad ones.' Counter-memes, to use Godwin's terms, have always operated in conjunction to debunk information or cultural ideas. As such, memetic battles, as it were, have been vehicles for political and social debate. This suggests similarities with the aforementioned cellular structure of power, which social media and memes being used to construct and consolidate political discourse.

The anthropomorphic Pepe the Frog is a popular meme, which, in its most extreme reiterations, has been used as a symbol of White nationalism and the alt-right.[12] Originally published as a cartoon strip by Matt Furie for his 2005 comic series *Boys Club*, Pepe is a green frog with a humanoid body, exaggerated white eyes and green lips (although in some depictions his lips are red). Pepe's laid-back disposition and wide grin was accompanied with the catchphrase 'feels good man.' On bodybuilding forums in 2008, users began posting images of themselves post-workout with Pepe's feels-good caption, crossing over to use on websites such as 4chan and Reddit[13] as a reaction image[14] to workout content (Pelletier-Gagnon and Trujillo Diniz, 2021). By 2009, Pepe had developed traction on numerous websites, and over the years went through a series of alterations to constitute a meme, including 'Sad Frog,' with tearful eyes and a frown, with the new catchphrase 'feels bad man.' This was used on 4chan bodybuilding forums by young men countering toxic standards of masculinity as physical fitness (Logan-Young, 2021). Pepe was no longer only used to celebrate muscle gain and weight loss, but instead to convey feelings of dissatisfaction or societal pressure. This demonstrates early subversions of the original meme to contradict its initial meaning; a counter-meme, as Godwin described.

'Smug Pepe' became popular during Donald Trump's 2016 Republican presidential campaign. Trump shared the image on his X account (formerly Twitter) along with a caption that read: 'You Can't Stump the Trump.' Pepe is positioned behind a podium, wearing a blue suit, red tie, and white shirt, with his blonde hair (or toupee) styled into a combover. This is a clear reference to Trump's caricature. One of Pepe's green hands is resting on his chin, and while his expression appears to be one of thoughtfulness, there is an undertone of malice to his sly grin. Trump's Democratic opponent, Hillary Clinton, condemned the image, and her official campaign website denounced Trump sharing the Pepe the Frog meme as promoting symbols of White supremacy and hatred (Kozlowska, 2016). This demonstrates the use of internet memes in campaigning strategies and Pepe's popularity forced it to be discussed in American presidential elections (Denisova, 2019). Trump's affiliation with the far right, including his accepting of KKK David Duke's endorsement, and calling for a ban on all Muslims entering the US (Chan, 2016; Anderson, 2016) solidified the association of Pepe to the alt-right. The meme has been used by other right-wing advocates to promote

Trump's racist policies on immigration and the policing of the US–Mexico border. For example, one meme featured Smug Pepe (representing Trump) standing in front of a fence with a US border sign, in reference to Trump's anti-immigration wall. Behind the fence is a man wearing a poncho and sombrero, and a woman holding a baby to represent Mexican migrants. The man is grabbing the fence with an expression of anger, while Smug Pepe slyly grins at the viewer as though he celebrates their obstruction (Toor, 2017). The journey that Pepe the Frog has made from 'feels good man' to a symbol of the alt-right shows that memes do not have an inherent meaning. They are shaped by the individuals who engage with them and are embedded in their social context.

Whilst early memes took the format of static images, due to technological advancement, along with the proliferation of platforms, such as TikTok, they have now developed into videos. A popular meme on TikTok is the use of the 'Bearded Cutie' filter, which gives women the appearance of facial hair while they mimic alpha podcasters. Although the creators and content are varied, they all use the same filter to parody the rhetoric of alpha podcasters. There is a notable linguistic attribute to memes by reappropriating speech; whether a catchphrase, citing a political opponent (on any spectrum) or podcaster. In its new, memetic context, this same speech elicits humour. Speier (1998), who will be further discussed in relation to humour, refers to 'migratory' political jokes, which are sustained across different contexts with minor adaptations, such as language or geopolitical references. The meme may loosely be considered as sharing some characteristics of a migratory joke given that it changes in circumstance by nature of being individually reproduced, whilst its central comedic intent is consistent. In the parodying of alpha podcasts and the Bearded Filter cutie, the underlining meme is the speech of the alpha men themselves. Effectively, by 'migrating' the same misogynist speech, but in a clearly sarcastic context, they dismantle its credibility. For example, Elsa Lakew who uses the Bearded Cutie filter told NBC news:

> It feels like a new video clip was going viral every other week of some guy spewing some garbage take on women ... I got pretty sick of it. And so, when I saw that filter on TikTok and used it, I immediately thought of these podcasts guys ... Since we couldn't fight logic with illogical takes, parodying them was the next best thing.
>
> *(Sung, 2022)*

Kimber Springs and Lilly Brown are TikTok users who produce content parodying alpha podcasters, creating the spoof, *Fresh New Tits*, a title clearly mocking *Fresh & Fit*. Springs and Brown imitate alpha podcasters by using the same speech, but as spoken by and adapted to the opposite gender. That is, they refer to 'high value, alpha women' instead of men.

As alpha podcasters claim women are duty bound to provide sex to men, Springs and Brown respond that men must provide emotional intimacy (highlighting and parodying the trope of women being over-emotional). As alpha podcasters cite their high value in fitness and wealth, Springs and Brown cite homemaking: 'I cook good food, I have a spotless house, I make money. What are you bringing besides maybe $40,000 a year and a whole lot of headache?' (@kimbersprings, 2022a). As alpha podcasters bemoan the loss of a woman's value by her sexual partners, Springs and Brown retort 'as males grow older, their natural hairline recedes, and they lose their value' (@kimbersprings, 2022b). They subvert misogynist discourse by mimicking their language and dislocating its signifying chains by reframing it from a woman's perspective (albeit sarcastic), thus making humour of something violent and oppressive and, quite literally, laughing at misogyny. In one video (@kimbersprings, 2022c), they invite a guest, whom they are at pains to describe as 'male' in reference to the alpha podcaster's emphasis on 'females.' They overlay music as soon as he begins to speak, in direct comparison to sexist podcasters audible obstruction of women's voices. Many other women TikTok users produce similar content, though not necessarily reaching the engagement of Springs and Brown. At the time of writing, the *Fresh New Tits* podcast videos had amassed over 3.5 million views, 1.6 million likes, 30,000 comments, and 40,000 shares. This demonstrates its 'viral' capacity, receiving a wide audience and engagement, the perceived success of posts measured by this so-called clickable revenue.

Laughter as Refusal

Speier (1998) categorises jokes as aggressive, defensive, diversionary/ soothing, or cynical. Whilst the former may be directed at individuals, or in scoring a political victory, the cynical joke is more generalised whilst still operating at an ideological level. In highlighting the characteristics of making something funny via, for example, surprise, nonsense,[15] puns, or the omission or transposition of letters (spoonerism), the TikTok counter-memes exemplify the use of defects – applying characteristics such as stupidity to ridicule political or public figures. This may refer to intellectual or bodily defects, such as the mockery of alpha podcasters as balding, often directed at Gaines, who is recently rumoured to have undergone a hair transplant. Political jokes are acts of communication between the joker, the person who laughs, and the 'victim' or butt of the joke, revealing the relative power between them. Accordingly, jokes may be between equals or elicit the laughter *of* the powerful. However, and most relevant to this discussion, they may also induce the laughter of the powerless, boldest when a joke is made at the expense of the powerful, which 'in their presence' may be 'highly dangerous' (Speier, 1998: 1392). Whilst not necessarily at a physical proximity, the

TikTok memetic joke indeed utilises the same social media space as the alpha podcasters it mocks. Such 'danger' may not be an in-person threat, but can manifest through cyberbullying and online harassment. In extreme cases, it can lead to 'swatting' or 'doxxing,' whereby rivals prank-call the authorities to prompt an armed response by SWAT teams, or leak home addresses of popular figures online. This has certainly led to real-life, physical consequences, including death from heart attacks (Cramer, 2021) or being shot by police (Helmore, 2018). An infamous example of harassing women commentators online is Anita Sarkeesian, a feminist critic of sexist tropes in video games who was bombarded with death, rape, and bomb threats, and a game was created within which players could punch and visibly bruise her avatar (Webber, 2017; Krendel, 2022). Whilst studies of social media have included consideration of its ability to transcend the regulation of public space by dictatorial regimes[16] (Matsilele and Mututwa, 2021), in a gendered context, it may also be utilised by women and girls to mitigate the physical and sexual dominance of the misogynist men they undermine, though clearly, a corresponding online threat remains.

Feminist theorist, Bonnie Honig's (2021) work on refusal discusses the use of sound via chanting, which could both excite a crowd to frenzy, but also be used to pacify, to calm, and unite. She discusses the role of the chorus, which we compare to the collective sound-creation of memes and the corresponding laughter, as a forum for women to share and unite; 'if a single chorus whispers tragedy, then three together may just sing jubilee. If a single chorus only ventilates desire to serve the marriage plot, three together may upend it' (Honig, 2021: 90). By collectivising their voice, the song, or the meme, becomes a unifying force. Speier notes that whilst laughter forges ties between people, it 'does not change the circumstances that it illuminates, although it is able to lessen the discontent and even the despair that these circumstances produce' (1998: 1358). Thus, the singular act of a meme or laughing is not a revolutionary act in isolation, but on a mass scale, its sound can reverberate against hegemonic discourse. Discourse is not limited to spoken language but includes visual signifiers, like the use of facial filters. Memes are therefore discursive in nature and as per their defining trait of being reproduced and adapted by multiple participants, have a particular propensity to discursive ruptures. Their humour is directly harnessed by subverting the context of speech, images, or public figures, applying numerous other texts or imagined realities and images to it, thus making their audience laugh who recognise something has been displaced and reorganised. Although humour does not destroy power, it caricaturises it (Speier, 1998). When oppressed groups laugh at dominant individuals or structures, their power is challenged as though the subjugated group refuses to conform (Bhungalia, 2020). Internet memes in particular produce an unconventional space for political participation with everyday citizens as its primary creators and audience, which, aside from making people laugh, enables them to engage with different perspectives (Tsakona and Popa, 2013). As George Orwell (1945) famously remarked, 'every joke is a tiny revolution' because it 'upsets the established order' and 'brings down the mighty from their seats, preferably with a bump.' This 'lingual politics' enables participants to speak over and across dominant discourse, to realise new subjectivities (LaBelle, 2018). In this case, women and girls disrupt the credibility of alpha podcasters, and rather than being subservient and absent from social media, they are loud and laughing. Such laughter, which cannot entirely collapse power relations, nonetheless calls attention to it.

Despite the dismissal of 'feminist killjoys,' feminists have long employed humour for political ends and to bring attention to sexism in media. For example, the 1993 Barbie Liberation prank, whereby the voice boxes of 50 Barbie dolls were swapped with G.I Joe dolls to highlight gender stereotyping in children's toys, or the 2016 protest by Women in Film and Television who attended a red carpet dressed as sausages, demanding more inclusion of women in the film industry (French, 2023). Such acts refuse, then, to be silent, to conform, to take seriously sexist norms. In theories of humour, these constitute jokes of incongruity. That is, humour based on the strange or the abnormal, which provokes laughter and pleasure at the nonsensical (Mpofu, 2021; Kant, 1790). In the above cases, the ridiculousness of attributing gender stereotypes to inanimate objects, or the visualisation of women dressed as phalli.

Sound has a coercive quality in so far it can compel us to respond both physiologically and psychologically (LaBelle, 2018). Laughter is an instinctive response to something heard or seen, a joyful and involuntary performance of listening. As LaBelle (2018) explores sonic agency as a means by which new conceptualisations of the public and its membership emerge, so too laughter may consolidate political communities, which are formed in the spontaneous sharing of laughter. In other words, 'unlikely publics,' which may not take the form of large movements but manifest in day-to-day relationships, potentially contradicting master narratives. We suggest a *memetic public* is formed through the creation and sharing of memes, which, though utilising an online space, may expand and influence gender politics on a wider, macro level. Feminist campaigner and filmmaker, Janey Stephenson (2017) argues that 'the fight back need not be joyless … holding onto and creating joy is a vital tactic for personal and group resilience, as well as political resistance to an agenda that seeks to enforce hierarchy.' An example which parallels the Bearded Cutie filter to some extent, is drag. Judith Butler (2011) writes of drag's deconstruction of gender through humour, by bodies 'appropriating' (in non-exploitative terms), their contrary assigned gender and/or class, which by insisting on inclusivity ('serving realness' for example, meaning to convincingly portray another identity), subsequently parodies the exclusivity of gender binaries. Drag Kings are funny because they dislocate the ownership of masculinity. By 'appropriating' the speech of alpha podcasters, women on TikTok make visible its exclusive privileging of men which, when spoken by the out-group (women and girls), becomes laughable. This inclusion/exclusion is further served in laughter, on the one hand signalling being *in* on the joke whilst marking social groups with which one may or may not belong. For example, laughing together may reveal an alignment of social experience, cultivating group solidarity. Therefore, women laughing together can consolidate political communities, which can influence and agitate for change. As has been discussed, the use of online fora and memetic videos facilitates a sense of community across geographic distances and by co-participation in the joke, as exemplified by the comment section under Springs and Brown's videos.[17] Women and girls express their mutual recognition and disdain of online misogyny through the medium of TikTok. Against the hegemony of structural and online sexism, the deployment of feminist humour on TikTok, creating content *by* women *for* women, makes such laughter defiant.

Conclusion

We have argued that the production of memetic videos by women on TikTok and the subsequent laughter of audiences can be understood as ridiculing and opposing hegemonic masculinity. Although many creators may not explicitly state their content is feminist, they nonetheless challenge misogynist discourse popularised and consolidated on social media. As this chapter has shown, podcasts such as *Fresh & Fit* and *Tate Speech* have become a cause for concern not only due to their widespread consumption but the consequences this is having for young men and women in real life. By utilising Foucauldian concepts of power/resistance and discourse analysis, we argued that memetic videos created in response to alpha content offer opposition to patriarchal power cells. This obstructs the repetition of misogynist discourse and its normalisation. Women's memetic feminism directly contradicts and confronts alpha podcasters on their own turf. The creation of humorous counter-memes levels out the playing field of online discourse and encourages feminist discussion by inviting others to consider the violence of online misogyny, whilst being creative and accessible in their critiques. Non-traditional forms of feminist activism, such as meme creation, allows individuals that may not otherwise contribute (and particularly young people as those most targeted), to engage with socio-political issues. The struggle against misogyny is far from over. Whilst Andrew Tate and *Fresh & Fit*'s official platforms are deteriorating, their toxic ideology remains prevalent and must be challenged; perhaps, as it were, by any memes necessary.

Notes

1 Online spaces for men, typically used to describe sexist content.
2 Discourse refers to language and non-linguistic means by which certain social relations are produced, or forms of knowledge are claimed. In effect, discourse is the text and thought that *does* something to society and subjectivity. For example, discursive practices of business may include modes of speech alongside non-language components, such as dress or formal etiquette. These consolidate to produce multiple expressions of power, subsequently re-enforcing cells of power as we will later discuss.
3 In Real Life, a common abbreviation used online.
4 We have chosen to avoid citing this website so as not to indirectly promote the content.
5 A nodal point is a privileged sign, which acquires its meaning and stability through other signifying chains.
6 This is not exclusively used by men. Women too, participate in bolstering traditional views of femininity. For example, 'trad-wife' (traditional housewife) influencers (for example, The Radiant Wife and Mrs Midwest) combine an aesthetically pleasing, romanticised lifestyle with alt-right politics to engage their audiences (Zahay, 2022). Trad-wife influencers share sentiments with misogynistic podcasters, such as the need for women to embrace their 'natural femininity' to become more visually attractive to men. They also promote traditional ways of living to make women 'better' mothers.
7 The term was coined by R. W. Connell in the 1980s and has since been critiqued as based on heteronormative assumptions of masculinity. The full extent of these debates is not explored here; however, for a discussion on its origins and Connell's response to rebuttals, see Connell and Messerschmidt (2005).
8 Fan pages not run by the public figure themselves. 'Copycat' refers to imitating or being influenced by another.
9 Li and Blommaert (2020) discuss a similar phenomenon in China with women deliberately provoking verbal abuse from a known misogynist internet personality, Liujishou. Sending him pictures of themselves, they baited him to insult their appearance and personality. Liujishou farmed out nonsense and ridiculous remarks, which were obviously baseless given he had no prior relationship with them, and failed to offend the women who had specifically requested to be insulted. Rather than being involuntary victims of hate speech, this draws parallel to women on TikTok who actively expose misogynist content to encourage widespread ridicule and laughter.
10 The term 'clap back' derives from African-American feminist linguistics and refers to quick-witted, direct rebuttals or insults, typically on social media (Washington, 2020).
11 We acknowledge concerns about Dawkins' own misogynist beliefs. These include blaming the sexual assault of women on their alcohol consumption and disparaging comments against those who campaign against sexual harassment. He has repeatedly defended his comments about paedophilia, claiming the fear surrounding the sexual abuse of children is overblown (Lee, 2014; Ohlheiser, 2013).
12 A contemporary and common term to describe the far right.
13 These websites are notorious members of the manosphere and have hosted some of the harassment levelled at Anita Sarkeesian, later discussed (Krendel, 2022).

14 A reaction image is similar to a gif or emoji, used to respond to posts and comments on social media.
15 Speier specifically refers to Dadaist poetry. For further discussion of Dadaism, see Chapter 3 ~~Dangerous~~ Dada?
16 For discussion on the use of public song to counteract the policing of public space, see Chapter 6 'Remove Them All!'
17 For example, the online acronym LOL meaning 'laugh out loud' is a written way to express laughter.

References

Anderson, M. (2016) Charts: Here's What Donald Trump Has Said On The Issues, *NPR*, 17 Nov, available at: www.npr.org/2016/11/17/501582824/charts-heres-what-donald-trump-has-said-on-the-issues. Accessed 3 Sep 2023.

Bhungalia, L. (2020) Laughing at Power: Humour, transgression and the politics of refusal in Palestine, *Environment and Planning, C. Politics and Space*, 38(3): 387–404.

Butler, J. (2011) *Bodies That Matter: On the Discursive Limits of Sex*. London and New York: Routledge Classics.

Ceci, L. (2022) Distribution of Global TikTok Creators 2022 by Gender, *Statista*, 7 Nov, available at: www.statista.com/statistics/1257710/tiktok-creators-by-gender-worldwide/#:~:text=Distribution%20of%20global%20TikTok%20creators%202022%2C%20by%20gender,app%20were%2055.3%20percent%20in%20the%20examined%20period. Accessed 21 March 2023.

Chan, M. (2016) Donald Trump Refuses to Condemn KKK, Disavow David Duke Endorsement, *Time*, 28 Feb, available at: https:// time.com/4240268/Donald-trump-kkk-david-duke/. Accessed 3 Sep 2013.

Connell, R.W. and Messerschmidt, J.W. (2005) Hegemonic Masculinity: Rethinking the Concept, *Gender & Society*, 19(6): 829–859.

Cramer, M. (2021) A Grandfather Died in 'Swatting' Over His Twitter Handle, Officials Say, *The New York Times*, 24 July, available at: www.nytimes.com/2021/07/24/us/mark-herring-swatting-tennessee.html. Accessed 10 July 2023.

Dahir, I. (2023) How to Spot an Andrew Tate Follower, *Buzzfeed News*, 20 Feb, available at: www.buzzfeednews.com/article/ikrd/andrew-tate-courses-pimpin-hoes-degree. Accessed 18 April 2023.

Das, S. (2022a) Andrew Tate: Money Making Scheme for Fans of 'Extreme Misogyny Closes', *Guardian*, 20 Aug, available at: www.theguardian.com/media/2022/aug/20/andrew-tate-money-making-scheme-for-fans-of-extreme-misogynist-closes. Accessed 10 March 2023.

Das, S. (2022b) Inside the Violence, Misogynistic World of TikTok's New Star, Andrew Tate, *Guardian*, 6 Aug, available at: www.theguardian.com/technology/2022/aug/06/andrew-tate-violent-misogynistic-world-of-tiktok-new-star. Accessed 17 Feb 2023.

Dawkins, R. (1976) *The Selfish Gene*. Oxford: Oxford University Press.

Denisova, A. (2019) *Internet Memes and Society: Social, Cultural, and Political Contexts*. New York: Routledge.

Divorce.com staff (2023) 5 Signs of Walk Away Wife Syndrome, *Divorce.com*, Last Updated 1 Mar, available at: https:// divorce.com/blog/walkaway-wife-syndrome/. Accessed 13 July 2023.

Dodgson, L. (2023) 'Fresh and Fit' Choked Up after Announcing it had been Demonetized on YouTube, *Insider*, 22 Aug, available at: www.insider.com/fresh-and-fit-host-cries-when-announcing-podcast-demonetization-2023-8. Accessed 25 Aug 2023.

Evans, A. (2023) Andrew Tate: How Schools are Tackling the Influence, *BBC News*, 11 Jan, available at: www.bbc.co.uk/news/education-64234568. Accessed 18 March 2023.

Farah, H. (2023) TikTok is the Most Popular News Source for 12 to 15-year-olds, says Ofcom, *Guardian*, 20 July, available at: www.theguardian.com/technology/2023/jul/20/tiktok-is-the-most-popular-news-source-for-12-to-15-year-olds-says-ofcom. Accessed 20 July 2023.

Foucault, M. (1976) [2020] *The History of Sexuality Volume 1: The Will to Knowledge*. London: Penguin Classics.

French, L. (2023) Protest is Dangerous, but Feminists have a Long History of using Humour, Pranks and Stunts to Promote their Message, *The Conversation*, 07 March 2023, available at: https://theconversation.com/protest-is-dangerous-but-feminists-have-a-long-history-of-using-humour-pranks-and-stunts-to-promote-their-message-199298. Accessed 11 Dec 2024.

Fresh & Fit (2021) Heated Debate! Are Women Entitled to a Man's Time w/o s*x? w/4 Girls, 14 July, available at: www.youtube.com/watch?v=j2u9C0QQYgw&t=3760s. Accessed 7 July 2023.

Glaze, V. (2022) Fresh and Fit Claim they're Victims of 'Targeted Attack' following YouTube Hate Speech Ban, *Dextero*, 6 Apr, available at: www.dexerto.com/entertainment/fresh-fit-claim-theyre-victims-of-targeted-attack-following-youtube-hate-speech-ban-1799476/. Accessed 15 March 2023.

Godwin, M. (1994) Meme, Counter-meme, *Wired*, 1 Oct, available at: https://www.wired.com/1994/10/godwin-if-2/. Accessed 1 September 2023.

Helmore, E. (2018) Tyler Barriss, Accused of Making Hoax Call, Regrets Death of 'Swatting' Victim, *Guardian*, 15 Jan, available at: www.theguardian.com/us-news/2018/jan/15/tyler-barriss-swatting-death-regret. Accessed 10 July 2023.

Hobbes, T. & Macpherson, C.B. (1981), *Leviathan*. Harmondsworth: Penguin.

Honig, B. (2021) *A Feminist Theory of Refusal*. Cambridge, Massachusetts: Harvard University Press.

Hunter, J. (2022) Is Andrew Tate Running a Pyramid Scheme? Hustler's University Exposed? *Ginx*, 21 Aug, available at: www.ginx.tv/en/tiktok/is-andrew-tate-hustlers-university-a-pyramid-scheme. Accessed 12 July 2023.

Källstig, A. and Death, C. (2021) Laughter, Resistance and Ambivalence in Trevor Noah's Stand-up Comedy: Returning Mimicry as Mockery, *Critical African Studies*, 13: 338–355.

Kant, I. (1790) [1987] *Critique of Judgment*. Indianapolis/Cambridge: Hackett Publishing Company.

Keller, J.M. (2012) Virual Feminisms: Girls' Blogging Communities, Feminist Activism, and Participatory Politics, *Information, Communication & Society*, 15(3): 429–447.

@kimbersprings (2022a) *What do Males bring to the Table of High-value Women?*, TikTok, 23 Feb, available at: www.tiktok.com/@kimbersprings/video/7056486813057469743. Accessed 21 Feb 2023.

@kimbersprngs (2022b) *Should Males make more Effort to keep their Hair?*, TikTok, 25 Jan, available at: www.tiktok.com/@kimbersprings/video/7056929209478073646. Accessed 21 Feb 2023.

@kimbersprings (2022c) *Allowing a Male to Speak on Our Podcast*, TikTok, 8 Feb, available at: www.tiktok.com/@kimbersprings/video/7062311585112509743. Accessed 21 Feb 2023.

Kozlowska, H. (2016) Hillary Clinton's Website now has an Explainer about a Frog that Recently became a Nazi, *Quartz*, 13 Sep, available at: qz.com/780663/hillary-clintons-website-now-has-an-explainer-about-pepe-the-frog-a-white-supremacist-symbol, Accessed 3 Sep 2023.

Krendel, A. (2022) From Sexism to Misogyny: Can Online Echo Chambers Stay Quarantined? In I. Zempi and J. Smith, *Misogny as Hate Crime*. Oxon and New York: Routledge, pp. 99–118.

LaBelle, B. (2018) *Sonic Agency: Sound and Emergent Forms of Resistance*. London: Goldsmiths Press.

Lee, A. (2014) Richard Dawkins has Lost it: Ignorant Sexism gives Atheists a Bad Name, *Guardian*, 18 Sep, available at: www.theguardian.com/commentisfree/2014/sep/18/richard-dawkins-sexist-atheists-bad-name. Accessed 31 Aug 2023.

Li, K & Blommaert, J. (2020) 'Please abuse me': Ludic-canivalesque Female Masochism on Sina Weibo, *Gender and Language*, 14(1), 28–48.

Logan-Young, F. (2021) Image Analysis: What happened to Pepe the Frog? *Different Route*, 21 Mar, available at: www.differentrooute.com/post/image-analysis-what-happened-to-pepe-the-frog. Accessed 23 Aug 2023.

Manne, K. (2018) *Down Girl: The Logic of Misogyny*. New York: Oxford University Press.

Mathers, M. (2023) School holds Andrew Tate Assemblies over Fears of 'Toxic Misogynistic' Influence on Children, *The Independent*, 11 Jan, available at: www.independent.co.uk/news/uk/home-news/andrew-tate-social-media-gateway-drug-b2260309.html. Accessed 17 Feb 2023.

Matsilele, T. and Mututwa, W.T (2021) The Aesthetics of 'Laughing at Power' in an African Cybersphere. In S. Mpofu. *The Politics of Laughter in the Social Media Age: Perspectives from the Global South*. Switzerland: Palgrave Millan, pp. 23–42.

Mpofu, S. (2021) Ridicule and Humour in the Global South: Theorizing Politics of Laughter in the Social Media Age. In S. Mpofu. *The Politics of Laughter in the Social Media Age: Perspectives from the Global South*. Switzerland: Palgrave Millan, pp. 1–20.

Mustafa, F. (2022) Three time Andrew Tate's Comments on Women Enraged Social Media, *HITC*, 5 Aug, available at: www.hitc.com/en-gb/2022/08/05/three-times-andrew-tates-comments-on-women-enraged-social-media/. Accessed 15 March 2023.

Ohlheiser, A. (2013) Richard Dawkins Defends 'Mild' Paedophilia, Again and Again, *The Atlantic*, 10 Sep, available at: www.theatlantic.com/international/archive/2013/09/richard-dawkins-defends-mild-pedophilia-again-and-again/311230. Accessed 31 Aug 2023.

Oppenheim, M. (2023a) Three Women to sue Andrew Tate Over Alleged Sexual and Physical Assaults, *Independent*, 13 Apr, available at: www.independent.co.uk/news/uk/crime/andrew-tate-civil-legal-action-uk-b2318624.html. Accessed 13 April 2023.

Oppenheim, M. (2023b) Figures that Lay Bare the Shocking Scale of Toxic Influencer Andrew Tate's Reach among Young Men, *Independent*, 16 Feb, available at: www.independent.co.uk/news/uk/home-news/andrew-tate-influence-young-men-misogyny-b2283595.html. Accessed 17 Feb 2023.

Oppenheim, M. (2023c) TikTok 'Failing to Act' as Andrew Tate Videos Still Seen by Children as Young as 13, *Independent*, 2 July, available at: www.independent.co.uk/news/uk/home-news/andrew-tate-tristan-videos-tiktok-b2362496.html. Accessed 12 July 2023.

Orwell, G. (1945) Funny, but not Vulgar, *Leader Magazine*, 28: 1945.

Paul, K. (2022) Dangerous Misogynist Andrew Tate removed from Instagram and Facebook, *Guardian*, 19 Aug, available at: www.theguardian.com/us-news/2022/aug/19/andrew-tate-instagram-facebook-removed. Accessed 17 Oct 2022.

Pelletier- Gagnon, J. and Trujillo Diniz, A.P., (2021), Colonizing Pepe: Internet Memes as Cyberplaces, *Space and Culture*, 24(1): 4–18.

Plato, Lee, D. (2007) *The Republic*, 2nd edn, London: Penguin.

Playboard (2023) *Fresh and Fit*, last updated 30 Aug, available at: https://playboard.co/en/channel/UC5sqmi33b7l9kIYa0yASOm. Accessed: 15 March 2023.

Reid, C. (2022) Controversial Podcaster Claims Women having Instagram in a Relationship is Cheating, *Ladbible*, 20 Apr, available at: www.ladbible.com/news/podcaster-claims-women-having-Instagram-is-cheating-20220420. Accessed: 14 Nov 2022.

Rentschler, C.A. and Thrift, S.C. (2015) Doing Feminism in the Network: Networked Laughter and the 'Binders Full of Women' Meme, *Feminist Theory*, 16(3): 329–359.

Shrimsley, R. (2015) Cashing in on the Outrage Economy, *Financial Times*, 24 Apr, available at: www.ft.com/content/9d48e7b8-e92c-11e4-a71a-00144feab7de. Accessed 12 July 2023.

Shrivastava, A. (2022) How Many Followers does Andrew Tate have on Twitter? Controversial Personality Garnered Millions of Followers after Twitter Unban, *Sportskeeda*, 21 Nov, available at: www.sportskeeda.com/esports/how-many-followers-andrew-tate-twitter-controversial-personality-garners-millions-followers-twitter-unban. Accessed: 18 July 2023.

Speier, H. (1998) Wit and Politics: An Essay on Laughter and Power, *American Journal of Sociology*, 103: 1352–1401.

Stahl, C.C. & Literat, I. (2022) #GenZ on TikTok: The Collective Online Self-Portrait of the Social Media Generation, *Journal of Youth Studies*: 1–22.

Sung, M. (2022) TikTok Users turn 'Alpha Male' Podcasters into a Viral Joke. *NBC News*, 1 Feb, available at: www.nbcnews.com/pop-culture/viral/tiktok-users-turn-alpha-male-podcasters-viral-joke-rcna13999. Accessed 12 July 2023.

Thomas, T. (2023) Three Alleged Assault Victims Launch UK Civil Claim against Andrew Tate, *Guardian*, 12 Apr 2023, available at: www.theguardian.com/news/2023/apr/12/three-alleged-assault-victims-launch-uk-civil-claim-against-andrew-tate. Accessed 11 Dec 2024.

Toor, A. (2017) France's Alt-right has Turned Pepe the Frog into Pepe le Pen/After Trump's Election, Online Groups are using Familiar Memes to Troll Opponents and Build Support for Marie Le Pen, *The Verge*, 6 Feb, available at: www.theverge.com/2017/2/6/14522542/pepe-the-frog-france-le-pen-meme. Accessed 04 Sep 2023.

Tsakona, V. & Popa, D. (2013) Confronting Power with Laughter, *European Journal of Humour Research*, 1(2): 1–9.

Washington, A.R. (2020) 'Reclaiming my Time': Signifying, Reclamation and the Activist Strategies of Black Women's Language, *Gender and Language*, 14(4): 358–385.

Webber, J. (2017) Anita Sarkeesian: 'It's Frustrating to be known as the Woman who Survived #Gamergate', *Guardian*, 16 Oct, available at: www.theguardian.com/lifeandstyle/2017/oct/16/anita-sarkeesian-its-frustrating-to-be-known-as-the-woman-who-survived-gamergate. Accessed 10 July 2023.

Wright, G. & Murphy, M. (2022) Andrew Tate Detained in Romania over Rape and Human Trafficking Case, *BBC News*, 30 Dec, available at: www.bbc.co.uk/news/world-europe-64122628. Accessed 15 Feb 2023.

Zahay, M.L. (2022) What "Real" Women Want: Alt-Right Feminity Vlogs as an Anti-Feminist Populsit Aesthetic, *Media and Communication (Lisboa)*, 10(4) 170–179.

5
ACOUSMATIC SOUND, NEOLIBERAL ANXIETY AND THEATRICAL RESISTANCE

Hara Topa

This is a story of resistance. A sonic resistance in theatrical performance. I query how acousmatic sound, is employed,[1] sound which has no visible source. I connect socio-political states of anxiety with the idea of an agitative effect inherent in acousmatic sound to show how it can be used by theatre directors and performance makers as a mode to resisting neoliberal economic oppression. I contend that, through the affective use of this acousmatic method, theatre audiences can not only be made aware of the effects of neoliberal oppression but can be inspired to effect social change. Theatre has a long history of effecting social change through awareness-building and inviting reflections on contemporary issues. Examples include Jeff Stetson's *The Meeting* (1987) about an imaginary meeting between Martin Luther King Jr. and Malcolm X that ignited discussions on civil rights and racial equality; Tony Kushner's *Angels in America* (1991), a play about the AIDS epidemic and its impact on the LGBTQ+ community in the 1980s, confronting audiences with issues of homophobia and discrimination. In the sections that follow, I examine two performances: the 2010 headphone theatre production *Cinemascope* by Blitz Theatre Group and the 2015 Dead Centre's headphone theatre production *Chekhov's First Play*. The audience in both performances use headphones, exemplifying the use of acousmatic sound. Both grapple with themes of devastation, which I interpret in a socio-economic context. Specifically, I read the performances in light of the 2008 recession, which later evolved into the European Debt Crisis. In analysing these performances, I explore the ways acousmatic sound is used both to stage and resist the impasse created by the neoliberal mode of perpetual crises (Hardt and Negri, 2017) and its residual anxiety. In other words, the standstill, the point where no further progress or development can be made. To portray neoliberal

DOI: 10.4324/9781003361046-6

oppression, I locate structures and figures of authority in the *acousmêtric* narrator of *Cinemascope* and Anton Chekhov's original play. The agitation and uncertainty of the faceless, neoliberal representation is further expressed by ventriloquist bodies whose sound (voice) comes from a source other than their own. By persisting in a state of dislocation and deadness, they operate as a prospective mode of resistance by turning/substituting their initial acousmatic utterances of desperation into/with voices of perseverance and protest.

My approach is interdisciplinary; acousmatic sound can be found in music theory – as is the antithetical de-acousmatisation (the point where the source of the sound becomes visible). The a*cousmêtre*, via film theorist Michel Chion (1999), can be traced to film theory and ventriloquism via psychoanalysis. I begin by introducing a brief history of the origin of the word 'acousmatic.' This reveals a connection between the nature of acousmatic sound and authoritative power. Such authoritative power refers to structures and discourse that exemplify extreme economic inequalities and the disingenuous facet of neoliberalism that promised democratisation and freedom but 'is quietly undoing basic elements of democracy' and replacing it with plutocracy, 'by and for the rich' (Brown, 2015: 7). I liken this to the 2009–2014 European Debt Crisis and its aftermath. It has personal significance to me as I have lived experience of the uncertainty and anxiety it engendered. Part of my sonic experience of economic oppression is available for you to listen in the QR code embedded below. This acousmatic narration portrays how my personal sonic environment changed when the banks in Greece shut down during the Debt Crisis. It was recorded in early July 2015. Greek citizens were only allowed to retrieve a small amount of money through ATMs (€60 a day) due to capital controls, some of which were enforced until September 2019. This led to noisy queues near ATMs that stretched for blocks with empty ATMs leading to complaints and, occasionally, violence.

History

In ancient Greece, the philosopher, Pythagoras, conceived of an intriguing way to keep his disciples focused on his teachings. During his lessons, he hid himself behind a curtain or veil, denying his students visual access to his person. He contended that the only way for his students to absorb his teachings adequately was to focus on the sound of his voice, devoid of any visual distractions. Brian Kane (2014) noted how Pythagoras' students, or *akousmatikoi* (listeners, from the Greek word *akousma* meaning the object of hearing), were obliged to stay behind the veil for five years before they were deemed worthy to listen to their master's lectures in full view.

However, this chapter is not concerned with ancient Greek philosophers: it is first and foremost about power and resistance. More specifically, the use of acousmatic sound as a theatrical tactic against neoliberal oppression. Typically, an oppressor must be identified as an object/subject to rebel against. The act of veiling in the Pythagorean example was ingrained with notions of power and authority. The use of the veil ensured that Pythagoras, the keeper of knowledge, was separated from his students, his indisputable authority weighing over the cohort. Failure to adhere to the rules of the master could result in being ousted from the congregation, disapproval, and likely ridicule. Pythagoras was viewed as a dominating figure; he was said to possess a superhuman or divine nature, performing wonders and leaving a legacy of the so-called Pythagorean[2] way of life, which was followed hundreds of years after his death. The inability to match the source of sound to an image can cause confusion, agitation, and insecurity. It is difficult to prepare yourself adequately to face, endure, or resist an authoritative power when you cannot be certain of the physical form the oppression takes.

The link between economic precarity and psychological anxiety is elucidated by political theorist, Mauricio Lazzarato, who claims that the purpose of neoliberal government is to 'construct memory, inscribe "guilt" in the mind and body, fear and "bad conscience" in the individual economic subject' (Lazzarato, 2011: 130). Polish philosopher and sociologist Zygmunt Bauman conceived the term 'liquid fear' (2006) to speak of similar uncertainty and anxiety under capitalist regimes. Neoliberal economic control exercises a continuous state of insecurity and uncertainty – i.e., the enforcement of austerity measures on several economically weak European countries. These measures created a politically violent culture of debt that incited protest and civil unrest, challenging public trust in institutions of power. Lazzarato (2012) described the creation of the 'indebted man,' a subject under economic oppression, subdued by the weight of his personal responsibility and guilt for an economic state imposed on them by a faceless organisation:

> The German press campaign against Greek parasites and loafers bears witness to the violence of the guilt engendered by the debt economy. When it comes to talking about debt, the media, politicians, and economists have only one message to communicate: 'You are at fault,' 'You are guilty'.
>
> *(Lazzarato, 2012: 31)*

In 2010, Greece was feared to default on its debt to the European Union, threatening the viability of the Union itself (Grexit). As a result, the European Union was forced to make additional loans to Greece, Portugal, Italy, Ireland, and Spain (attributed with the derogatory moniker PIIGS, coined in 2008 by Professor Andrew Clare). The high debt levels of these nations led to them being considered threats to the stability of the Eurozone and were faced with tough austerity measures. Economic policies dictated by the International Monetary Fund (IMF) demanded a severe decrease in wages and pensions while taxation was substantially increased across indebted European nations. The subsequent social turmoil, along with the instability caused by the fear of a potential collapse of the European Union, mark this period as one of vast political importance and indicative of great economic precarity. For many European citizens, the ensuing anxiety had no clear point of origin. The IMF stood as a faceless entity demanding obedience. Who created this economic upheaval? Was it individual governments? Europe? The IMF? The Greeks themselves? Greeks were criticised as parasites and loafers unable to effectively manage their economy. Their eventual punishment, in the form of debilitating sanctions, led to distrust of the institutions meant to economically protect the individual, and to the preservation of security and certainty. Lazzarato outlines the intentional continuum of crises and the structured disempowerment of individuals toward a perpetual state of fear.

> Current labour and 'workfare' policies ... introduce, to varying degrees, insecurity, instability, uncertainty, and economic and existential precarity into the life of individuals. They render insecure not only the lives of individuals but also their relationships with all the institutions that used to protect them ... there is a differential of fears that runs the entire length of the continuum.
>
> *(Lazzarato, 2017: 16)*

With plenty of blame, there was no distinct agent to attribute it to, other than faceless organisations and generic structures of authority. Subsequently, responsibility is placed on individual nations and citizens. I suggest that this effacement, the inability to identify economic perpetrators, is a deliberate extension of neoliberal oppression. Historian Quinn Slobodian (2018) notes how the global economy is portrayed as something unknowable, keeping the perpetrators of economic oppression faceless. To identify the sense of dislocation arising from such facelessness, I use the term 'neoliberal anxiety.' Neoliberal anxiety is marked by a precarious relationship of economically induced fear and control between governments or authorities and citizens that can only function in a condition of crisis. The management of crises 'is the mode of operation of neoliberal governance' (Hardt and Negri, 2017: 217). A state of crisis for example, legitimises the existence of particular authorities or institutions, and endows them with power and even force. Such state-sanctioned, and ultimately state-induced violence is what philosopher and cultural theorist, Slavoj Žižek, named 'objective violence' as 'the often catastrophic consequences of the smooth functioning of our economic and political systems' (Žižek, 2008: 1–2). He contrasts this systemic violence with 'subjective violence,' which has an identifiable agent. Consider a scenario within which an individual shoots civilians at a bank, enraged because he cannot afford his mortgage. This constitutes subjective violence, and the shooter can be identified as the source – the subject – of violence. The bank, exponentially raising the interest in mortgage payments is exercising systemic or 'objective' violence. Only the bank has no identifiable face. It's a faceless entity. The high mortgage payment order did not come from the teller or the bank manager. There is always a veiled distance obscuring the perpetrators. Such veiling may be deliberate in displacing accountability.

It's the End of the World as We Know It (and I *don't* feel fine)[3]

In 2010, in a small alleyway near Piraeus Street in Athens, the world ended in nine days. Nobody knew how or why. Just that it ended after a series of catastrophic incidents, threatening weather conditions, riots, and terrorist attacks. There were moments of insanity, cannibalism, and love. This is all known because, as the audience members find themselves taking their seats in the BIOS arts centre, the voice of an unknown narrator in their headphones makes sure that they know the details of the 'sad story of the end of the world' (*Cinemascope*, 2010). The audience have been warned by the same authoritative voice that the 'use of the headphones is obligatory, and any attempt to take them off is strictly forbidden' (ibid.). At some point, they will be given instructions that are imperative to follow. The voice carries a sense of solemnity. The audience is, after all, going to witness the end of the world, at least as intended by the Greek theatre group Blitz, through the headphone performance *Cinemascope*.[4]

It is in that very first encounter with this disjointed, bodiless voice of the narrator that I locate the authoritative figure; the voice commands, the audience obeys. The voice dictates how and where to move, establishing rules and prohibitions. The faceless narrator in this case draws parallels to institutions of power that decide the rules and demand obedience. Throughout the performance, he provides information on how the catastrophe unfolded. The bodiless narrator appears to know everything: the timeline, the events. The audience does not get the opportunity to slowly discover for themselves how events unfolded. They are being told the horrific details by the narrator who usurps the power of revelation.

The narrator's instructions extend over all the characters, demonstrating not only his familiarity with them but his control. He refers to them by their first names. His orders are followed without hesitation. He rudely and impatiently interrupts Franz, one of the characters wailing because he failed to save everyone, including the woman he loves, from the impending doom: 'Leave, Franz, now.' Franz immediately stops mid-sentence and hastily moves away in his wheelchair. The narrator questions another character about her feelings towards the end of the world, and dissatisfied by her reaction, demands that she cry, *now*. He speaks in past tense, as though standing above history, and detached, looks down to recount incidents distant to him. At the same time, he uses the first-person plural ('we were cold') aligning himself with the lives that perished in the catastrophe. The authoritative voice of the invisible narrator exists as the law, the wise, and ultimate. It insinuates that it might exist simultaneously in and beyond the catastrophe; the audience cannot know, they can only imagine.

> This day, the clouds went away, and a white light made its appearance. This light, which was called the Righteous Light, revealed all the flaws on the buildings and in the people. This was the justest [sic] day of all. We were cold. Never before was the city so beautiful.
> *(Cinemascope, 2010)*

To explore the supposed all-powerfulness of this disembodied voice, I draw on the notion of the *acousmêtre*, the cinematic figure of acousmatised sound as constructed by film theorist and experimental music composer, Michel Chion (1994). One of Chion's distinctive contributions is his exploration of sound in cinema, and the reciprocity of that relationship. Chion, being a music composer himself, found particular interest in Schaeffer's (1952) theorisation of music, especially in the notion of the 'sound object,' a sound unit complete and independent from the body it is produced by. This prompted him to investigate the intricate relationships created between voice and body in their cinematic manifestation.[5] In his seminal work, *The Voice in Cinema* (1999), Chion explores the haunting presence of acousmatic sounds in cinema, stressing its magnetic and disturbing effect on audiences. According to Chion, this effect is inherent in acousmatic sounds due to:

> Being in the screen and not, wandering the surface of the screen without entering it, the acousmêtre brings disequilibrium and tension. He invites

> the spectator to *go see, and he can be an invitation to the loss of the self, to desire and fascination.*
> *(Chion, 1999: 24, original emphasis)*

Chion imbues the acousmatic with a supernatural, almost godly power: 'the greatest *Acousmêtre* is God' (Chion, 2011: 164). His views on acousmatic sound have been enduring, further developed by a number of subsequent theorists. Artist and theorist, Brandon LaBelle speaks of the disturbing effect of the disembodied voice:

> The acousmêtric voice, by its nature, catches our attention by surprise. When this uncanny sonic percept enters experience, the listener is suddenly struck by a voice that appears to come from nowhere. When sound is dislodged from its source our ability to single out its originating location is adversely affected. Consequently, thematic attention often becomes fragmented and dispersed.
> *(LaBelle cited in Home-Cook, 2015: 92)*

Chion identifies a number of different types of *acousmêtre*s; complete, already visualised, commentator, radio, even a theatre-*acousmêtre* in which he suggests that 'the offstage voice is clearly heard coming from another space than the stage – it's literally located elsewhere' (Chion, 1999: 22). Chion intentionally left the definition of the *acousmêtre* open to avoid an impermeable definition that would exclude other manifestations. This allowed for consideration of theatrical performance, especially in regards to issues of authority and narration. Authority in the *acousmêtre* is located in its presentation by Chion as an all-powerful (and thus potentially controlling and menacing) entity. He refers to Alfred Hitchcock's elusive mother in *Psycho* (1960) and the super-computer, Hal 9000 in Stanley Kubrick's *2001: A Space Odyssey* (1968).

> First, the *acousmêtre* has the power of seeing all; second, the power of omniscience; and third, the omnipotence to act on the situation. Let us add that in many cases there is also a gift of ubiquity—the acousmêtre seems to be able to be anywhere he or she wishes.
> *(Chion, 1994: 129–30)*

Chion's acousmêtric characteristics are evident in *Cinemascope*'s bodiless narrator. The narrator is in the audience's ears, he commands the performers, he resides everywhere and nowhere at the same time, the missing body invisibly occupying many spaces, commanding all of them. He is heard in the foyer giving instructions. He is there while the audience is being seated in the performance space in front of a glass screen resembling a shop window. He is in the alleyway we see through the window and the buildings across it, weaving, and sometimes directing, the narrative of the last nine days of the world. The 13 performers in the alleyway outside the window, accompanied by unsuspecting passers-by, follow his instructions. In view of such a powerful *acousmêtre*, one might doubt whether the authority of the acousmêtric voice could be contested at all. Chion (1999) argues that the power of acousmatic sound would dissipate if the acousmêtric voice was revealed to occupy a visible body. This is the point of de-acousmatisation. A classic example is the man behind the curtain in *The Wizard of Oz*. When the curtain drops, the supposed all powerful wizard is revealed to be simply a man operating controls. However, *Cinemascope*'s acousmêtric narrator is never revealed. The bodily source of the sound remains hidden to the very end, rendering the narrator a complete *acousmêtre* in Chion's terms.

Writing seven years post-Chion, philosopher and cultural theorist, Mladen Dolar (2006), argues that an intrinsic characteristic of the voice is its in-between-ness: between the physical body and language. Dolar subsequently rejects de-acousmatisation as impossible:

> There is no voice without a body, but yet again this relation is full of pitfalls: it seems that the voice pertains to the wrong body, or doesn't fit the body at all, or disjoints the body from which it emanates ... The fact that we see the aperture does not demystify the voice; on the contrary, it enhances the enigma.
>
> *(Dolar, 2006: 70)*

Dolar's notion of the wrong or misfit body[6] finds application in the 13 performers of *Cinemascope*, their voices often muffled and distorted by overlapping sound effects and utterances by other performers. The sound of their voices are intermingled with other instances of acousmatic sound such as non-distinctive voices over radio, repetitive, crackling gunshots, helicopter sounds, muffled cries, thunder, and explosion sound effects. Sometimes the viewers see the performers opening their mouths, mouthing off into the air desperately, but no sound is discerned. There is a very specific anxiety that ensues as the audience must imagine the missing sound. And yet, a voice that does not fit the bodies in view fills the headphones: the narrator. It belongs to a missing body – one that the viewers search to identify behind the glass but fail.

This sense of not belonging, of displacement, and its resulting anxiety causes distress to the viewer, a state best described by theatre scholar, Pieter Verstraete, as 'auditory distress ... an *excess* of intensities in the listener' (Verstraete, 2011: 83, original emphasis). This excess, the multiplicity of voices and sounds, latching and extracting from the mismatched body, is what Verstraete calls 'radical vocality,' a congregation of vocal sounds in a performance. Verstraete recognises disembodiment as a prominent characteristic of radical vocality, which he equates with acousmatic sound and the ventriloquist voice (Connor, 2000). Verstraete proposes that radical vocality is more than just an incoherent or disturbing sound. Rather, he places emphasis on its meaning-making abilities, especially in the context of theatrical performance. Due to its alienating quality, the disembodied voice in performance challenges perceptions and reveals the 'potential to resist hegemonic ways of perceiving and understanding' (Verstraete, 2011: 86). Audience members are able to perceive the disturbing aspect of acousmatic sound and draw connections to the actions on stage. The intentionality of the performers' dislocated vocality can potentially lead the audience to seek the reasoning behind the unnatural dislocation. The ensuing discomfort and agitation warns of danger, of something not being right, and invites the audience to form a different understanding of what is happening on stage, or search for an alternative meaning. An audience at the time of economic oppression could visualise their current socioeconomic state of consecutive crises in the performance and seek to assign a different meaning than that of surrendering to the inevitability of debt. The passivity and regularity of following the oppressive mandates of faceless institutions of power can be challenged through the uncomfortable acousmatic agitation and ignite thoughts of resistance to economic oppression. The audience, therefore, may consider vocalising their dissent, actively express their need for a change in their circumstances and consequently enforce social change (Figure 5.1).

FIGURE 5.1   Syllas Tzoumerkas as Franz and Marialena Mamareli as the caretaker in *Cinemascope*. © *Georgios Makkas*, courtesy of Yorgos Valais.

The Dead Ventriloquist Body

Often, the bodies behind the *Cinemascope* glass appear doll-like. They resemble puppets, their movements vigorous, mouths open but without voice. This is essentially a performance of ventriloquism.

> Every emission of the voice is by its very essence ventriloquism. Ventriloquism pertains to voice as such, to its inherently acousmatic character: the voice comes from inside the body, the belly, the stomach—from something incompatible with and irreducible to the activity of the mouth.
> *(Dolar, 2006: 79)*

Dolar is building on Žižek's (2001) exploration of a lack of stable embodiment:

> An unbridgeable gap separates forever a human body from 'its' voice. The voice displays a spectral autonomy, it never quite belongs to the body we see, so that even when we see a living person talking, there is always a minimum of ventriloquism at work: it is as if the speaker's own voice hollows him out and, in a sense, speaks 'by itself,' through him.
> *(Žižek, 2001: 58)*

Žižek is preoccupied with ideology from a Marxist perspective. He comprehends ideology as a set of discourses, which may be employed to justify economic relations and profit. According to philosopher, Louis Althusser (1970), who influenced Žižek, ideology is not just conscious speech but embedded through the arts, culture, education. Althusser stresses that individuals do not construct their ideologies in a vacuum. They do not make decisions on which ideologies to follow from a detached and objective point of view. They are always influenced by the ideologies that constitute the societies they inhabit; from discourses adopted in the close environment of family and friends to those governing education, media, governments. Žižek (2001) explains that this leads to a false perception of political reality, limited to the confines of existing political frameworks. He advocates for challenging these realities. 'Formal' freedom operates within existing power relations, whilst 'actual' freedom undermines these very dynamics (Žižek, 2001: 122). When individuals shed the constructed ideologies that plague them and reformulate oppressive discourse, this discursive site of intervention is one of 'actual' freedom, or at least creates a point of resistance. Audiences watching the *Cinemascope* performance are being exposed to that intervention point. The agitative qualities of ventriloquist bodies and sound dislocate audiences from their numbness and disrupt normality. In this way *Cinemascope* makes room for a point of actual freedom. There is a point between spaces of formal and actual freedom, a space between constant crises and social change, a liminal space ripe for resistance, the space between life as it is and catastrophe. This space, or gap will be referred to as an impasse. Where ideology fails, where the cause of catastrophe is unknown, there is only a nihilistic space full of dislocated voices. Those voices are breaching the hull of their presumed bodies. *Cinemascope*'s premise is that, by the end of the performance, the bodies the audience encounters will have already met their demise. Not only are they ventriloquised through acousmatisation, but they are ventriloquised by the fact that their bodies will soon no longer be alive. They are anxiety-ridden entities, impossible to comprehend in the acousmatic dislocation of their voices. By the close of the performance, the audience becomes aware that the voices heard belonged to people eviscerated in this unidentified catastrophe.

Cinemascope was first performed in 2010, during severe austerity measures devastating the Greek economy, which had already been suffering since 2008. I therefore read the impending catastrophe and loss of life in the performance as mirroring the anxiety-ridden reality of a society that staggers through what appears to be never-ending economic sanctions. These sanctions were experienced as imposed by a distant and faceless organization of power, with no possibility of future improvement. This devastating Greek reality can be seen in the space *Cinemascope* occupies and the exploration of how people respond to catastrophe. It is the end of times, the space of annihilation where life and death exist simultaneously as the performers are forced to continue living under the certainty of their soon-to-be evisceration. Social interactions, public affairs, and political processes become futile, facing the certainty of annihilation. The voice of an operator is heard through the audience's headphones, conducting what appears as a typical but ultimately futile poll. They can see a body of a woman holding a phone. The woman, however, is too far away to clearly distinguish if the voice belongs to her. It might seem logical to assume it is her but still, one cannot be certain. She announces she intends to take her own life: 'Do you feel we deserve what's been happening lately? Yes? No? Maybe? Ok. … I am thinking of committing suicide in five days' (*Cinemascope*, 2010). This signals an end in the utility of institutions; she enacts the only autonomy she has, albeit a fatal one.

The relationship of acousmaticity and the dead body is not a novel idea. It epitomised Chion's notion of the acousmêtre in 'the cinema, the voice of the acousmêtre is frequently the voice of one who is dead' (Chion, 1999: 46). There are several examples of dead narrators in film and television. Kevin Spacey's narrator in *American Beauty* (1999) and the popular television show *Desperate Housewives* (2004–2012) feature dead narrators. Chion explicates a state of undeadness in the cinematic acousmêtre: 'what could be more natural in a film than a dead person continuing to speak as a bodiless voice, wandering about the surface of the screen? Contrary to Chion's notion of the omnipotent acousmêtre however, the dead ventriloquist bodies in

Cinemascope initially appear devoid of power. They emphasise the emptiness of power when the end of the world is looming; revealing their inability or lack of total control. Questions arise as to the supposedly omniscient acousmêtric voice of the narrator as well; is he dead too? How powerful can the acousmêtric narrator be if there is no world left to exert his power over? It appears void, distressing, and inexplicably confusing as to how the assertions of control and management can be true if the world is nonetheless doomed.

The acousmatic nature of the dead bodies of the performers does more than confirm and convey neoliberal anxiety and fear. I contend that the acousmatic voice of the dead turns on itself, an impasse created, which implodes to resist its own annihilation. My claim borrows from Foucault's (1978) notion of the counter-conduct; a mode of resistance that may employ the same tools used to oppress. It works within the confines and boundaries of oppressive authorities (formal freedom) and is often a creative and indirect form of resistance. A simple example is a schoolboy forced to wear a specific uniform by the school he attends. His counter-conduct as a way of resistance would be to tie his tie differently than advised by the school to show protest. In 2011, the 'we won't pay' movement was flourishing in economically devastated Greece. Revolting against debilitating taxes and salary cuts, many Greeks decided to simply stop paying what they deemed unfairly imposed. Examples of counter-conduct include chaining electricity meters, ignoring toll booths and covering bus ticket machines with plastic bags.

Literary scholar, Steven Connor identifies, among other expressions of vocality, the voice of rage. According to him, the voice of rage, or seductive voice, 'is aimed at transcending its own condition, forming itself as a kind of projectile, a piercing, invading weapon, in order to penetrate, disintegrate and abandon itself' (Connor, 2000: 37). This can be heard in the voices of the 13 performers outside the glass front, heard through our headphones, performing a panic-stricken polemic of desperation. Franz's sobbing and desperate shouting about his failure to save the world, the anxious voice of a reporter confirming riots are raging throughout Europe. These form an acousmatic spear of vocality that disperses, folds, and then turns on its own desperation to become resistance. Even though it appears ineffectual in preventing the end of the world, it surpasses desperation by, as Connor puts it, achieving another affectivity in the performance. It gives rise to performative acts of resistance, such as the tragicomic lip-syncing of *Gloria in Excelsis Deo* in *Cinemascope*. The lip-syncing subject as an 'active agent' (Riszko, 2017: 160), resists its imposed and irreversible demise. It reverses the meaning of the hymn. Once originally attributed glory to God, when sung by the ventriloquist bodies of the performers, it ridicules the supposedly godly-like powers of the now destroyed authoritative structures. It is sung by ventriloquist bodies that persist to exist against annihilation, being their own authority figures in this impasse or in-between state. A man arriving on a Vespa dressed in a Superman costume announces that things can be fixed, shouting and out-of-breath; Superman will save the world because he always does. A woman using a megaphone states her anger and sadness, despairing how much time people wasted in life. Thus, the final scene of *Cinemascope* is not one of explosions and screams as one might expect of an apocalypse, rather it is one of complaint and lingering hope for a changed future. Rather than resigning themselves to obliteration, they recall the trials and tribulations of life as though it could shape the tomorrow, which may never come. The voices persist in a state between life and death, both foreshadowing and experiencing their eventual demise. They are the voices of the dead, after all, they are deathly.

> The power of the voice derives from its capacity to charge, to vivify, to relay and amplify energy. But, precisely because of this, the voice can also become deathly; ... a voice that is pure discharge, a giving out of mere dead matter, toneless, vacant, absent, sepulchral, inhuman. It seems to demonstrate that it has no connection with the world, or with the one who originates it.
>
> *(Connor, 2000: 38)*

The quasi-dead ventriloquist bodies, alluding to famished bodies, exist in the liminality of imminent extinction. There is almost something irreverent, almost impious, about their persistence to persevere in this state of certain death. The deathly, or, according to Connor, excremental voice can be a means of asserting one's existence or presence in the absence of meaningful communication, a form of resistance or refusal to engage in conventional forms of communication. There is no meaningful communication when the end of the world is approaching. The voice of authorities cannot survive the chaos of the catastrophe any more than its citizens. The instructions from the megaphone are confusing and devoid of power. The ventriloquist performers acousmatically assert their existence in the meaninglessness of a world that is about to end; they persevere against and resist the terrible fate wreaked by institutions of power.

> What the scream tears apart, it also holds together. The scream is the guarantee that, after the world has been atomised, it will reassemble and again resemble itself.
>
> *(Connor, 2000: 34)*

Cinemascope pertains to an intriguing paradox: the coalescence of rage and excremental voices. There lies the sustained and repeated rupture indicative of the aforementioned impasse, where the acousmatic voice of the dead resists its own deadness as a theatrical method. That is, they refuse to concede to their impending annihilation at the end of the world. This suggests a shift in systems of power. The acousmatic, dead, ventriloquist body ceases to simply portray the effect of neoliberalism and instead represents resistance as persistence. Franz's wailing is overpowered by a love song played through speakers. Desperation itself, falls apart.

Quasi-ventriloquism and Cruel Optimism

Dead Centre's *Chekhov's First Play* premiered in 2015 at Samuel Beckett Centre, Dublin. Anton Chekhov (1860–1904) was a Russian playwright, and in this adaptation of Chekhov's text, Dead Centre subverts his *Play Without a Title* also known as *Platonov*. *Chekhov's First Play* features themes of enforced debt crises and economic practices devastating Ireland since 2008/2009, producing social unrest. Chekhov's original story, set in a dilapidated country house, grapples with themes of failed ideals and disillusionment. These resemble the disappointment and frustration felt by Irish people due to economic adversity and precarity brought forth by faceless economic institutions. Contrary to Blitz' *Cinemascope*, the voice of the narrator in this headphone performance corresponds to a visualised body. Bush Moukarzel, the narrator, introduces the performance, testing the function of the headsets by repeating words and asking the audience if they can hear him. According to the narrator, previous audiences had not quite grasped the meaning of the play, so his commentary is required to help clarify the subtle nuances previously missed. There is no sense of omnipotence. The voice of the narrator in our ears is sympathetically received although, at times, it appears patronising in assuming the audience will not understand the performance. More crucially, the narrator is standing in front of us – nothing mystical about him. In this performance, economically oppressive structures of authority are not located in the narrator. The first act begins as originally intended by Chekhov, with the actors engaging in conversation on the stage, moving about or sitting at a table. However, the narrator has now disappeared, and the audience can only hear his voice. His newly acousmatic voice not only overlaps with the performers', but vehemently criticises their speech. Eventually, the audience realise that the timing of the dialogue is off. The performing bodies delay their vocal utterances. They cannot proceed until the disjointed, acousmatised voice of the narrator has finished speaking. They are dependent on it. They do not control the performance; the acousmétric voice does. The narrator in this case appears to assume authority but for a very different cause.

The acousmatic narrator dismantles Chekhov's play by cutting scenes and characters claiming the storylines were too complicated or the characters not to his liking. Like *Cinemascope*, the voice of the narrator becomes seemingly omnipotent. In the middle of the performance, having received instructions through his headphones, an audience member walks up to the stage and joins the performers. At this point, the narrator's voice falls silent. Pre-recorded voices of the performers read assigned texts, and the performers begin lip-syncing them. This emergence of the ventriloquist body functions differently in Dead Centre's performance. By demanding a deviation from the script and instructing the performers to destroy the set and take off their costumes, the narrator's voice appears to discard the authoritative power of the script and demand a new, energised direction. The narrator here is what Chion would have called an already visualised acousmêtre. Like a teacher who has imparted his wisdom, he falls silent while his 'students' (the other performers) are empowered to exhibit acousmatic resistance on their own. In this new, antithetical variation of Chekhov's play, the performers eventually do not submit to the fate dictated by the old script. They forge their own. They choose not to be overcome by economic desperation and succumb to the fate of the victim.

Many characters in Chekhov's four most important plays (*Cherry Orchard*, *Three Sisters*, *Uncle Vanya*, and *The Seagull*) feature characters that face suffering, disappointment, loss, unfulfilled dreams, and unrequited love. In Dead Centre's version, the characters escape that fate. The acousmatic voice of the narrator stresses the need for escape in what cultural theorist Lauren Berlant calls cruel optimism. In other words, 'the ordinary as an impasse shaped by crisis in which people find themselves developing skills for adjusting to newly proliferating pressures' (Berlant, 2011: 8). While *Cinemascope* conceptualises permanent crises through the instigation of anxiety, Berlant describes how consecutive crises become the norm, the ordinary. She explains that 'crisis ordinariness' ensues wherein 'a person or a world finds itself bound to a situation of profound threat that is, at the same time, profoundly confirming' (Berlant, 2011: 2). This situation places people in a terrible state, where they become accustomed to the debilitating effects of crises. In cruel optimism, a state of political or economic precarity may be sustained and endured by conforming subjects, under the comforting auspices of established structures and relations, no matter how threatening they might be. People who have spent a long time living in economically oppressive conditions have come to consider this ordinary life (as Žižek's theory of political reality attests). Thus, the ordinariness of the neoliberal state of perpetual crises, is simultaneously cruel and comforting. Cruel in the precarity it sustains, comforting in the habitual adaptation of the individuals accustomed to it. It forms an impasse similar to *Cinemascope*, a gap or paradox between devastation and the undying perseverance of individuals.

It is this ordinariness and cruelty that the ventriloquist bodies of *Chekhov's First Play* dismantle through acousmatic sound. In the first part of the performance, the ventriloquist bodies attempt to sustain Chekhov's classical play, continuing amidst numerous script cuts and changes in scenes. The performers are still trapped in Berlant's cruel ordinariness. The narrator, however, intends to disturb this conformity. His acousmatic voice informs us that 'in the original there was a whole storyline about a bank but it was too complicated, so I cut it' (*Chekhov's First Play*, 2015). The performers are forced to perform a broken script; lines and whole scenes are missing; the characters are not as they should be; the narrator scolds an actress who is pregnant because the character she is portraying is not. Chekhov's play has been fractured, it cannot be resumed as it was originally conceived. Yet, the bodies keep on performing. Cracks begin to appear; performers mispronounce names and lines, forgetting their script and in panic, stand muted, mouths frantically searching for the right words to say. Sound eludes the performing body – Dolar's misfit body – still the performing bodies persist, in a hopeless attempt to return to ordinariness. In *Cinemascope*, the neoliberal impasse and agitation was portrayed through deadness. The dead body is conveyed in Berlant's notion of ordinariness:

> crisis rhetoric belies the constitutive point that slow death … is neither a state of exception nor the opposite, mere banality, but a domain where an upsetting scene of living is revealed to be interwoven with ordinary.
> *(Berlant, 2011: 102)*

A state of deadness is therefore prevalent in Dead Centre's representation of perpetual neoliberal crisis and Berlant's ordinariness critique. In *Chekhov's First Play*, slow death underlies the performers' bodies as they initially try to be faithful to the original script within the 'broken' play. It is the narrator's quasi-ventriloquist voice that signals the beginning of resistance acts: 'I just haven't been feeling myself lately. And by lately, I mean ever' (*Chekhov's First Play*, 2015). This utterance, slightly altered, signifies the inevitable break from the old script and is later carried by all characters through lip-syncing, commencing vocal acts of resistance that appear to lead to a disengagement from cruel optimism. Acousmatic sound turns against its former conformity and now exudes resistance through change. The performers themselves, shedding their costumes, symbolically shed their prescribed characters as dead, evolving and infusing new life to them. The voiceover of Anna Petrovna is heard saying: 'I am running away with you Platonov. And I don't have to pay my student debt in the bank because I won't be living in Ireland anymore' (*Chekhov's First Play*, 2015), with another voiceover of multiple characters' voices finally coming to the painful realisation: 'Is this mine? I cannot imagine owning anything. You made me nobody' (ibid., 2015). The acousmatic agitation performed by the continuous lip syncing and voiceovers acts as an affective act of resistance. The performers shatter everything that sustained the ordinariness of perpetual neoliberal crises, dismantling and taking control of the previously prescribed Chekhovian dialogue.

Deadness is used as a theatrical activist tactic. Theatrical space hosts what Berlant (2011) calls an 'intimate public.' I propose that ventriloquism, as an inherent characteristic of the acousmatic voice, can lead to collective acts of resistance: 'in an intimate public ... collective mediation through narration and audition might provide some routes out of the impasse and the struggle of the present' (Berlant, 2011: 226). The ventriloquist voice in Dead Centre's performance reimagines the people who have been obscured under the crisis ordinariness as dislodged entities. Considering LaBelle's (2018) problematisation of the relationship between power and visibility, sound can provide agency for the obscured: 'Practices of acousmatic listening and invisibilities ... give challenge to the often insistent ways in which political acts and public life are understood by way of appearance' (LaBelle, 2018: 54–55).

This may feel familiar to an audience feeling the effects of economic devastation, experiencing a permanent state of disappearance under the economic oppression of faceless institutions of power. Envisioning the performing body as the body of the collectively oppressed under the normalisation of neoliberal practices of control, it is nevertheless the ventriloquist dead body that resists them. Acousmaticity as resistance gives power to oppressed individuals and restores their visibility.

A game of Russian Roulette in *Chekhov's First Play* poses the question: should the ventriloquist body die in order for the characters to escape unbearable living under perpetual crisis? The ventriloquist voice needs a (vocalic) body, although by definition this may not be their own:

> The vocalic body is the idea – which can take the form of dream, fantasy, ideal, theological doctrine or hallucination – of a surrogate or secondary body, a projection of a new way of having or being a body, formed and sustained out of the autonomous operations of the voice.
>
> *(Connor, 2000: 35)*

The narrator appears in the last minutes of the performance with a blood wound on his head suggesting his participation in the game. But the Russian Roulette does not kill the performers. We perceive the trickery; the ventriloquist voice is not silenced when the narrator's lips stop moving and we realise we have been hearing a recording, nor when he turns the gun on himself. It might never have belonged to him in the first place. The voice finds a new vocalic body to occupy. The aforementioned audience member who joined the performance at the outset, slowly exits the stage and utters his one and only line in the play: 'hello.' The man from the audience is sent back to the audience as a symbol of survival. To show the audience it can be done and is therefore worth persevering for.

Epilogue (It's the End of the World as We Know It and I feel fine)

It's done. The world has ended, Anna Petrovna has run away with Platonov. No debt was paid, and many ventriloquist bodies formed a resistance force in *Chekhov's First Play* against economic oppression. Franz is probably still sobbing in *Cinemascope*. I employed acousmatic sound to show how it can portray economic oppression through the authoritative power of the narrator in *Cinemascope* or the vocalisation of Chekhov's original text in *Chekhov's First Play*. I examined how acousmatic sound can turn on itself to signify resistance through the deadness of the ventriloquist bodies of performers. Acousmatic sound occupied the impasse that was imposed upon it and used it to voice resistance through perseverance in *Cinemascope* and rejection of slow death in *Chekhov's First Play*. Economic crises still happen – smaller, bigger. Neoliberalism marches forth. All is well. We are all fine. In thinking of social change, I proposed acousmatic sound can be a theatrical method of resistance to neoliberal oppression irrespective of the directors' intent. Theatre will do what it always does; affect, entertain, create new meaning, agitate, shed light to social issues, provide a voice to the marginalised, awaken, bore, ignite controversy, and so on. I do not presume that theatre can change the world, but in some cases, it has the potential to facilitate social change, especially through the seed of agitation carried within acousmatic sound.

Notes

1 An example of acousmatic sound in activism is the use of megaphones on mass demonstrations. Whilst they amplify voices, these are not always seen or able to be connected to specific bodies; rather the origin of any singular voice is hidden within the crowd. See also Chapter 6 for an example of the acousmatic in football stadiums.
2 This included religious rituals, dietary restrictions, and observing silence as moral discipline.
3 In 1987, the popular music group REM released its apocalyptic hit song, *It's the End of the World as We Know It (And I Feel Fine)*. Recounting the end of the world through hurricanes and earthquakes, Michael Stipe's voice reassures us that he feels fine.
4 For an extended trailer of Cinemascope with English subtitles, see the Blitz Theatre Group YouTube channel.
5 I note however, that Ross Brown (2020) considers that the notion of the acousmêtre can be applied to any sonic object whose source has been obscured.
6 Many theorists have tackled the complicated relationship between voice and embodiment, including Adriana Cavarero in her seminal work *For More than*

One Voice (2005). Cavarero's politics of the voice focus on the uniqueness of the voice and the rewrite of philosophical avenues through a feminist revisiting of female histories and mythologies.

References

Althusser, L. (1970) [1971]. Ideology and Ideological State Apparatuses (Notes towards an Investigation). In L. Althusser. *Lenin and Philosophy and Other Essays*. New York and London: Monthly Review Press.

Berlant, L. (2011). *Cruel Optimism*. Durham and London: Duke University Press.

Brown, R. (2020). *Sound Effect: The Theatre We Hear*. London: Bloomsbury Methuen Drama, an imprint of Bloomsbury Publishing Plc.

Brown, W. (2015). *Undoing the Demos: Neoliberalism's Stealth Revolution*. New York: Zone Books.

Chion, M. (1994). *Audio-Vision: Sound on Screen*. Translated from French by C. Gorbman. New York: Columbia University Press.

Chion, M. (1999). *The Voice in Cinema*. Translated from French by C. Gorbman, New York: Columbia University Press.

Chion, M. (2011). The Acousmêtre. In Corrigan, T. et al. (eds.). *Critical Visions in Film Theory: Classic and Contemporary Readings*. Boston: Bedford St. Martin's, pp. 156–165.

Connor, S. (2000). *Dumbstruck – A Cultural History of Ventriloquism*. New York: Oxford University Press.

Dolar, M. (2006). *A Voice and Nothing More*. Cambridge: MIT Press.

Foucault, M. (1978) [2007]. *Security, Territory, Population: Lectures at the College de France, 1977–78*. Basingstoke Palgrave Macmillan.

Hardt, M., & Negri, A. (2017). *Assembly*. New York: Oxford University Press.

Home-Cook, G. (2015). *Theatre and Aural Attention: Stretching Ourselves*. Houndmills, Basingstoke, Hampshire: Palgrave Macmillan.

Kane, B. (2014). *Sound Unseen: Acousmatic Sound in Theory and Practice*. New York: Oxford University Press.

LaBelle, B. (2018). *Sonic Agency: Sound and Emergent Forms of Resistance*. London: Goldsmiths Press.

Lazzarato, M. (2012). *The Making of the Indebted Man: An Essay on the Neoliberal Condition*. Cambridge: Mass, Semiotext(e).

Lazzarato, M. (2017). *Experimental Politics: Work, Welfare, and Creativity in The Neoliberal Age*. Cambridge, MA: MIT Press.

Riszko, L. (2017). Breaching Bodily Boundaries: Posthuman (Dis)Embodiment and Ecstatic Speech in Lip-synch Performances by Boychild. *International Journal of Performance Arts and Digital Media*, 13(2), pp. 153–169.

Schaeffer, P. [1952] (2012). *In Search of a Concrete Music*. Translated from French by C. North and J. Dack. Berkeley: University of California Press.

Slobodian, Q. (2018). *Globalists: The End of Empire and the Birth of Neoliberalism*. Cambridge, Massachusetts: Harvard University Press.

Verstraete, P. (2011). Radical Vocality, Auditory Distress and Disembodied Voice: The Resolution of the Voice-Body in The Wooster Group's *La Didone*. In Kendrick, L. and Roesner, D. (eds) *Theatre Noise: The Sound of Performance*. Newcastle upon Tyne: Cambridge Scholars Publishing, pp. 82–96.

Žižek, S. (2001). *On Belief*. London: Routledge.

Žižek, S. (2008). *Violence*. London: Profile.

6
'REMOVE THEM ALL!'
Sounds of Protest in the Algerian Hirak Movement

Stephen Wilford

In February 2019 public demonstrations began to take place across Algeria following an announcement that the 81-year-old incumbent President, Abdelaziz Bouteflika, would seek a fifth term in office. Bouteflika had long capitalised upon his role in fighting for Algerian independence from French colonial rule, his subsequent part in the coup d'état that had removed the autocratic President Ahmed Ben Bella, and his successes in bringing relative peace to the country after the bloody violence of the civil conflict known as the *Fitna*, which had lasted throughout the 1990s.[1,2] By 2019, however, Bouteflika had become increasingly associated in the minds of the public with the alleged corruption of *le pouvoir*, a secretive cabal of wealthy and influential politicians, military generals, and business leaders.[3] He was also wheelchair bound and virtually unable to speak after a series of strokes had left him incapacitated, and for many this further demonstrated that the real political power in Algeria no longer rested in the hands of an elected President. Growing discontent at issues of unemployment, housing, and economic inequalities had been simmering under the surface of Algerian society for decades, and these finally came to a head in 2019, with the incredulity of Algerian citizens from across the social spectrum quickly evolving into anger. Journalists and activists suggest that on the 22nd of February an estimated 800,000 people marched through the streets of the capital city Algiers in protest against Bouteflika, *le pouvoir*, and the entire Algerian political system (Berkani, 2019; Meddi, 2019). As anger grew and the protests continued, the public's demands coalesced into *Hirak*, a democratic movement calling for political change and social justice.[4]

A few weeks later, on the 11th of March, it was announced that Bouteflika would no longer be standing in the election, and that the ruling FLN (*Front de libération nationale*) party would be proposing a new candidate.⁵ For many onlookers, both within and outside of the country, it was presumed that the apparent successes of the protestors would result in the rapid demise of the *Hirak* and a return to the political status quo. However, for those who had marched through the streets of towns and cities across Algeria (and, in many cases, in cities outside of the country that contained a large Algerian diaspora, particularly in France), the situation was not so simple. The removal of Bouteflika, people suggested, would not cause any serious damage to the power and influence of *le pouvoir*, nor ensure that Algeria became a more socially equitable country. As such, it was quickly apparent that the emergence of the *Hirak* represented a much longer-term sense of injustice and anger shared by large swathes of Algerian society, and that the protests would continue until real social and political change had taken place.

Ytnahaw ga': Voicing Dissent

On the evening of the 11th of March 2019 an Algerian female Sky News Arabic reporter, Yasmine Moussous, was reporting live from the streets of Algiers. As cars drove past, their horns and radios blaring, and people shouted from open windows, Moussous proclaimed that Algerians were in a celebratory mood, stating that 'Algerians are congratulating each other for what has been accomplished thus far' (Bentahar, 2021: 476). Sofiane Bakir Torki, a young man employed in a local pizza restaurant, happened to be passing and, upon hearing this, interrupted the live broadcast, angrily interjecting that,

> We are not convinced at all. For change, they removed one pawn and put in place another pawn. Remove them all! Remove them altogether!⁶

After an initial attempt to silence Torki, Moussous decided to engage him in conversation. The journalist, herself Algerian, responded in *Fusha* (or Modern Standard Arabic), imploring Torki to speak 'Arabic.'⁷ The demand was immediately significant to Algerian and non-Algerian viewers alike, bringing forth centuries of debate and discord around language. Torki was speaking the local vernacular form of Arabic known as *Darija*, which is used widely throughout Algeria, Morocco, and Tunisia, and draws together vocabulary and grammatical structures from classical and Modern Standard Arabic, Tamazight, French, Spanish, and Ottoman Turkish. The diversity of linguistic sources from which *Darija* draws is symbolic of Algeria's history of occupation and cultural encounter, and the country's position as the nexus between Africa and southern Europe. The power struggles that have characterised Algerian politics and society throughout the colonial and postcolonial periods have often been best articulated through the relationship that certain groups within society have to language. While independence brought a desire to replace French with Arabic as the country's official language, and a political drive towards Arabization throughout Algeria, French has continued to be part of the education system and is spoken widely as a second language. *Fusha* is employed primarily for official purposes, with the majority of daily conversations taking place in *Darija* or one of the country's varieties of Tamazight. Writing about the role of language in postcolonial Algeria, Mohamed Benrabah (2013) suggests that 'dialectal Arabic and Berber would be minorized and stigmatized, Literary Arabic confined to the devotional sphere and traditional values, and French to more "prestigious" functions' (2013: 50).

Importantly, *Darija* is almost unintelligible to speakers of Modern Standard Arabic, and anyone from outside of North Africa. As such, the attempts of Moussous to encourage Torki to speak *Fusha* can be understood as a desire to ensure that he was understood by Sky News Arabic viewers. Nevertheless, Torki remained defiant and unwilling to adapt, declaring that 'I don't know Arabic, this is our *Darija*!'. Ziad Bentahar (2021: 484) situates the incident within the long and complex history of language within Algeria, arguing that,

> After years of a rigidly pan-Arab official narrative that did not reflect the realities of people's identities, Algerians stand proud before the world's eyes for the first time in over a generation. The *ytnahaw ga*' slogan indicates that language is also relevant to the *Hirak*. Where the turgid Modern Standard Arabic of corrupt moribund officials was undignified, the eloquence of *Darija*, in contrast, is something that the people can be proud of.

Encapsulated within this encounter were local and transnational discourses of class, language, education, wealth, and national identity. This resonated simultaneously with Algerian viewers and Arabic-speakers beyond the nation's borders. The encounter played out through the sonic, via the voices of the protagonists and the broader soundscape of urban Algiers at this particular moment. These scenes were captured by the microphones of the Sky News cameras. As such, the demand to 'remove them all' went far beyond the mouths and ears of those involved in this impassioned encounter. In the recording of the interview Torki and Moussous are forced to almost shout, their words nearly drowned out by the noise of passing traffic. This soundscape incorporated multiple sonic signs open to interpretation by viewers and listeners. Were the noises of car horns and the shouting emanating from vehicles representative of collective celebration or anger, or perhaps a mixture of the two? Did the images and sounds appearing onscreen suggest, as Moussous had insinuated, that Algerian citizens had achieved their aims, or rather, as Torki was claiming, did they represent disappointment and frustration?

The clip of the encounter has circulated extensively online and been turned into countless memes. Website comment sections and social media platforms have produced lengthy discussion and debate, with users mainly lauding Torki for his ability to voice the tensions that the *Hirak* has unveiled, and express the difficulties faced by Algerians throughout a number of previous decades. The significance that it was a young, working-class man from an inner-city neighbourhood responsible for creating one of the *Hirak*'s slogans has not gone unnoticed and has been described as symbolising the movement's ability to engage citizens across the social spectrum.

Theoretical Beginnings

This chapter aims to critically interrogate the role that music and sound have played in the *Hirak* movement, and the ways in which they have shaped sonic practices and socio-political discourse that engage with various contexts, media, and social structures. The intention here is to consider how and why the songs and sounds of the *Hirak* have come into being, and critique the ways in which they have circulated, within and beyond Algeria. An underlying assertion throughout the chapter is that music and sound have not only exposed and confronted the extant layers of power, agency, and censorship that prevail within contemporary Algerian society, but have also afforded people (musicians, listeners, protestors, and ordinary citizens) a means for expressing their anger, sadness, and frustration. As a result, the argument is put forward that music and sound have been intimately bound up in processes of protest and grassroots rebellion, providing a means of making audible a collective desire to produce a new form of political regime and civil society within the country.

The theoretical grounding of this chapter is shaped by Ben Tausig's (2018) notion of the 'sonic vernacular,' a term that he proposes for understanding the ways in which music and sound elucidate and configure localised forms of protest and dissent. As Tausig argues, protests do not simply emerge from nothing, and 'sound and hearing's role in political dissent is shaped over the course of vernacular histories' (2018: 16). Tausig goes on to describe 'sonic practices that make symbolic and performative sense where they are employed, but perhaps not elsewhere' (2018: 26). Such an approach is useful here because it enables us to appreciate the *Hirak* as something distinctive, a product of a particular set of political and social circumstances that are entirely unique to Algeria. This is important because it encourages us to recognise the agency of the Algerian public, while forcing us to recognise the movement as not simply an extension, or reimagination, of the so-called

'Arab Spring' and other anti-authoritarian actions throughout North Africa and the Arab world. Jill Jarvis (2021: 10), in her excellent analysis of Algerian literature through a decolonising framework, argues that,

> To observers persuaded by the version of decolonization invented by France, this peaceful uprising might look like a potential step forward in a teleological progression that has been best modelled by European democracies. However, the dignity revolution (*thawrat al-karāma*) taking place in Algeria cannot be described as simply a popular revolt against the dictator of a failed African state or another episode in a so-called Arab Spring. Such tropes and terms reflect distorting Eurocentric assumptions. Seen from the standpoint of the people who have put their bodies on the streets every week for the past year, this movement is a much more radical and powerful collective dispute with the cartographic and temporal frames that underwrite the coloniality of power itself. If we take seriously what many of the protestors themselves are saying, the *Ḥirāk* is the unfinished liberation war.

Jarvis's words point towards a need to understand the *Hirak* on the terms of Algerian citizens themselves, as a continuation of political action aimed at liberation from oppression. If we situate the *Hirak* within contemporary Algerian history, we might suggest that, despite hopes at the time, national independence in 1962 failed to afford equality of opportunity to Algerians or to produce social justice after more than a century of French colonial rule. The *Hirak* therefore remains an intensely Algerian movement, concerned with a specific set of historical and social conditions. Furthermore, while many of the songs and chants of the *Hirak* draw upon musical styles and traditions that circulate globally, not least hip hop, they are made meaningful in specifically Algerian ways, and for specifically Algerian audiences.

Situating the *Hirak* within Algerian History

The *Hirak* emerged in early 2019 but was a culmination of factors that had shaped Algerian politics and society throughout the colonial and, in particular, postcolonial periods of the country's history. The story of the Algerian anti-colonial struggle is a long and complex one, the details of which fall well beyond the parameters of a chapter of this length. It suffices to note that by the time independence was won in 1962 the dominant organisation spearheading the movement were the FLN. They initially formed as a political collective, they directed anti-colonial armed resistance throughout the mid-twentieth century and, in the form of a political party, assumed power when independence was won. Entering the relative instability of a postcolonial political landscape, the party has never been entirely unified, with different factions of the FLN vying for presidential power. The FLN has remained in power since 1962 and opposition candidates continue to struggle to attract the necessary number of votes to successfully provide a political challenge. As the connections between politicians, the military and business leaders have become increasingly blurred, the gap between the wealthiest and poorest within Algerian society has grown.

The *Hirak* is, in part, a response to the growing socio-economic inequalities that have become increasingly evident in Algeria, a country rich in oil and natural gas reserves. Lala Muradova (2016: 68)[8] suggests that Algerian political instability in recent years results from,

> its aging and ailing President (over 70 years old); power struggles between clans; the country's weak economy and heavy dependence on hydrocarbon exports; youth unemployment reaching almost 22 per cent; social injustice; entrenched corruption; endemic inequality; and the sense of *al-hogra* among its citizens.

The realities of life for many Algerians stand in stark contrast to the egalitarian socialist principles espoused by the political regime. Following independence, the FLN government foregrounded ideas of collective nationalism, venerating the struggles and sacrifices of the anti-colonial movement in what Martin Evans (2012) terms the 'one million martyrs narrative' (2012: 335).[9] The state simultaneously propagated a policy of Arabization, intended to replace French with Modern Standard Arabic as the national language. Not only did this solidify social stratification in failing to recognise the importance of *Darija*, as we have seen earlier, but it also ignored the status of other languages in the country, and in particular the *Tamazight* language used by local Amazigh populations. Miriam Lowi (2010: 177) argues that,

> The leadership, in its efforts at fashioning a nation and as part of the post-independence developmentalist agenda, espoused an ideology and rhetoric of popular incorporation, and implemented policies – such as heavy industrialization, arabization, and the distribution of abandoned colonial property – that it insisted were inclusive in nature. However, not only did those policies fail to achieve their stated goals, but they were, in practice, highly exclusionary.

Algerian identity after independence in 1962, at least according to the official narrative, is assembled around an idea of monocultural, Arabic-speaking, Muslim homogeneity. This belies the realities of Algerian society, both in the past and the present.[10] Alongside cultural and linguistic anxieties, issues of class, housing, and wealth distribution have shaped political action in postcolonial Algeria. Yet the voices of protest have frequently been suppressed and censored. However, as I will show in the following discussion, the *Hirak* has begun to challenge and deconstruct some of Algeria's social structures, and musicians and activists from across the social spectrum have become involved in sonic responses to these issues.

From the Football Terrace to the Street

Songs and sounds of protest in Algeria have a long and complex history. They encompass anti-colonial musical expressions from the period of French rule, and more recently, politically engaged hip hop that has drawn attention to the racism and discrimination experienced by North African diaspora communities in France. When, in 1980, demonstrations took place in the predominantly Amazigh region of Kabylia (*Tamurt n Leqbayel* in the local *Taqbaylit* language) calling for official recognition of Amazigh culture in Algeria, singers were at the heart of the protests. Discussing the 'Berber Spring' (*Tafsut Imaziyen*) in 1981 (p32), an article in the Index on Censorship noted that,

> The protests and demonstrations in Algeria's Kabylia region, which ended with violent clashes and riots in Tizi Ouzou in April 1980, were notable for the leading role taken by musicians and poets. Two well-known Kabyle singers, Ait Menguillat and Ferhat, were arrested (they have now been released), and local groups were prevented from performing and on occasion detained by the authorities. Despite the widespread unrest accompanying these acts, which certainly took the Algerian government by surprise, such discrimination is not an isolated occurrence, but reflects long-term official discrimination against Berber cultural life.

A decade later, extreme violence materialised on the streets of Algeria following the cancellation of national elections in which the Islamist opposition party, the *Front Islamique du Salut* (FIS), looked likely to win a majority and take power. Armed anti-government groups, such as the *Groupe Islamique Armé* (GIA), waged war on the government and targeted journalists, artists, and musicians. A number of well-known musicians were killed over the ensuing decade, including the Kabyle singer Matoub Lounès. Debates about who was responsible for his murder continue to this day, with accusations levelled at both the Islamist groups and the government, as Lounès was a vocal critic of both sides during the *Fitna*. Christopher Schaefer (2015: 31) writes that,

Matoub advocated for secularism, freedom of speech, and for Kabyle linguistic and cultural rights. He did so unreservedly and with great courage in some of the most innovative music and beautiful poetry to have ever emerged from the Djurdjura Mountains. In the defence of his beliefs and his community, he fearlessly tread on sacred nationalist ideology and sacred religious beliefs. Those attacks on sacred beliefs, be they religious or national and linguistic, ultimately cost him his life.

Musicians such as Matoub, who have been willing to engage in outspoken criticism of the government, however, remain outliers within postcolonial Algerian history. Condemnation of the state has often been considered a taboo subject, at least within a public context, and heavy-handed policing practices have ensured that dissent has been kept to a minimum. One public space has provided an arena for the composition and performance of anti-authoritarian protest songs: the terraces of the country's football stadia. Football is the country's national sport and garners significant interest, particularly among Algeria's working-class male population. And the stadia in which matches are played offer fans a rare sense of collective safety from the threat of violence or censorship from the state. Throughout matches songs that openly critique the authorities are performed, and Nassim Bella (2019) reflects that,

> Inside the stadiums, where the young, the rebellious, and the discriminated against could meet, a degree of free space was possible to express collective rejection and opposition to the regime. Here, the ultras excelled in creating slogans and composing songs often very hostile to the regime; yet, thanks to the degree of solidarity among fans, stadiums provided a temporary "safe zone" that the police forces rarely dared to enter.

The football terrace therefore functions as both a public and quasi-private safe space, in which the individual is sheltered while the collective becomes both visible and audible. This shift, from personal vulnerability to collaborative safety, I argue, is fundamental to understanding the agency that football songs and chants have provided to Algerian citizens. The refuge that these concrete terraces provide, and the close physical proximity that they produce, offer the ideal conditions for collective acts of musicking that express ideas and opinions that might be unspoken in other social contexts within Algeria.

The history of political protest song within Algerian football stadia can be traced back to the *Fitna*, when young people found themselves increasingly alienated from national politics and caught between the violence of the state and the armed Islamist opposition. In an attempt to shift their focus away from the conflict being fought throughout the country, many young people, and men in particular, turned inwards, concentrating their attention upon the football teams that represented their local neighbourhoods and cities. Mahfoud Amara (2012) suggests that in Algeria 'the sense of belonging of football fans is not necessarily constructed around the "sacred" tie with the nation, but rather regards the nostalgic vision of the neighbourhood (*houma*)' (2012: 53). In other words, football has reshaped and challenged the logics of national cohesion and homogeneity that the government has propagated since independence, placing a far greater emphasis upon local and regional identities. Assia Boundaoui (2015) adds that 'all around the country, young people became loyal to their local teams, and the football stadium became one of the only spaces where marginalised young people could feel a sense of belonging and freely assemble' (2015). Throughout the 1990s and early 2000s, songs took on a particular sense of importance for football fans, with an increasing degree of organisation and, at least in some cases, significant musical aptitude. As Boundaoui (2015) writes,

> Football chanting is an art in Algeria, every club has its own distinct chants and song writers, its own drummers and ad hoc instrumentalists, and every football club in Algeria sings about overtly socio-political issues. The stadium is now the place where young Algerians can openly express themselves, shout their frustrations, declare their desires and sing about their dreams.

The football club most widely recognised for the political engagement of its supporters is USMA, the *Union sportive de la médina d'Alger*.[11] The club draws its support primarily from the working-class neighbourhoods of the capital city Algiers, and in particular from the historic Casbah in the heart of the city. Their red and black colours adorn walls and shop fronts around the local area, appearing alongside the green and red of their local rivals MCA, *Mouloudia Club d'Alger*, with whom they share the Omar Hamadi stadium. In 2010, the most committed of USMA's supporters formed an ultras group entitled *Ouled el Bahdja* (the Sons of Algiers) to facilitate and organise their musical activities, both within and beyond the stadium.[12] Their songs are more than just the typical terrace chant, often featuring a verse and chorus structure, and drawing upon a wide range of musical influences. A prominent source of musical inspiration is Algerian *chaabi*, an urban style that emerged in the early twentieth century as part of what Nadir Marouf (2002: 12) terms the 'indefinable nebula' of Algerian *andalusi* music. While *andalusi* has often been co-opted by the state as a symbol of Algerian cultural achievement throughout the postcolonial period, *chaabi* has offered a more accessible music that incorporates urban working-class musicians and listeners. Furthermore, *chaabi* emerged from the cafes of the Casbah of Algiers, the same streets from which USMA and MCA draw much of their support. Mickaël Correia (2019) writes that 'USMA supporters, mostly from the Casbah—the historic heart of Algiers and home of Algerian *chaabi*—are inspired by this traditional music, with its roots in Arab Andalus, and also by the social engagement of *raï*' (2019).

After the shock of the initial 22 February 2019 protests in Algiers, football matches throughout the country were cancelled by the authorities. The fierce local derby between USMA and MCA was allowed to go ahead on the 14th of March. The theories for why this decision was taken were numerous and often fuelled by collective suspicion. Did the authorities believe that the match would provide a distraction from the *Hirak* and encourage people away from the protests taking place on the streets of the city? Or might the match have been considered a way of encouraging politically-rebellious football supporters to come together within a single space? There were certainly fears that the match would lead to covert monitoring from the authorities and the potential for state violence to be directed towards the fans. Yet as many supporters stayed away from the match, often claiming that the aims of the *Hirak* were far more important than a game of football, the ultras of both USMA and MCA were in attendance, continuing to sing anti-government songs. As such, even at the height of uncertainty during the earliest days of the *Hirak*, these fans displayed their confidence that the football stadium could remain both a safe space for political expression, and a place for sonic resistance to the authorities.

While USMA's *Ouled el Bahdja* were known among the footballing community in Algeria prior to the *Hirak*, it was the release of a song in 2018 that would raise their profile both nationally and internationally. Entitled *La Casa del Mouradia*, the song was quickly adopted as an unofficial anthem of protestors as they marched through the centre of Algerian towns and cities. The title of the track is a playful pastiche that draws upon the Spanish Netflix series *La Casa de Papel* (entitled Money Heist in English), a story of a group of bank robberies, which has proven extremely popular with North African audiences. Ouled el Bahdja's song, however, also includes reference to the El Mouradia Palace, a grand building that sits in the hills above the centre of Algiers and which houses the office and official residence of the Algerian President. In employing the song's title to align a television programme about a series of bank robberies and the country's Presidential headquarters, the group's opinions of Bouteflika and their country's government are left in little doubt.

The song began life as a chant, performed a cappella by USMA supporters on the terraces of stadia across Algeria, before a studio recording was made by the members of the Ouled el Bahdja group, featuring multi-track vocals accompanied by a simple acoustic guitar. The recording was released across various media and streaming platforms in April 2018, including Spotify and

YouTube, which exposed the song to new audiences, particularly among the Algerian diaspora. However, with the lyrics of the song performed in *Darija*, *La Casa del Mouradia* remains almost entirely comprehensible only to Maghrebi listeners. The chorus of the song narrates the struggles and sense of despair experienced by many young Algerians, expressed in both the singular and plural first-person, while each of the four verses candidly critiques Bouteflika's previous terms in office.

Following regular performances at USMA's Omar Hamadi stadium and a growing online listenership, the song was adopted as an unofficial anthem during the early weeks of the *Hirak* movement. As protests began to be held weekly, not only in Algiers but in towns and cities across the country and in France, *La Casa del Mouradia* quickly became a soundtrack of urban protest and collective action. The song reverberated around urban spaces, broadcast from mobile phones, passing cars, and loudspeakers, embodying a form of communal expression that reflected the indignation of large swathes of Algerian society. When, under considerable pressure from protestors and political advisors, Bouteflika announced on the 11th of March that he would no longer be standing for re-election, it was *La Casa del Mouradia* that jubilant crowds sang outside the iconic Post Office building in central Algiers. A few hours later, the widely broadcast encounter between Sofiane Bakir Torki and Yasmine Moussous took place on a nearby street.

La Casa in the Concert Hall

The shift from football terrace to urban street, mediated through social media and online streaming services, saw *La Casa del Mouradia* enter a new public sphere, and one in which any form of political dissent has often been suppressed in postcolonial Algeria. The result was a general broadening of the song's listenership, from a primarily young male working-class audience in the country's football stadia to the much wider socio-economic spectrum that took part in the early *Hirak* protests. However, this was not the endpoint of the songs' journey of repositioning within Algerian society, and a further step saw *La Casa* infiltrate the more elitist worlds of Arab-Andalusi music and the concert hall.

Andalusi (alternatively Arab-Andalus, Andalouse, Andalusian, or Nūba) music is a term used to describe a variety of interlinked regional musical traditions that are performed throughout North Africa, and which purport to trace their history back to the Spanish Reconquista and the expulsion of 'Moors' (Muslims) from Iberia between the 9th and 15th centuries. The imposition of French colonial rule in Algeria throughout the 19th and 20th centuries involved the denigration, and often denial, of Algerian cultural identities, and when independence was finally won in 1962, Andalusi music became one of the symbols of Algerian national pride. Throughout the postcolonial period, Andalusi has often garnered interest and support from Algeria's educated middle classes, and, as Jonathan Shannon (2015: 47) writes,

> With independence, most North African nations made the Andalusian musical repertoire part of their national cultural patrimony. As a result, Andalusian repertoires were standardized and codified into official songbooks, and national music conservatories today serve as the primary context for learning the inherited performance practices.

During the summer of 2019, videos began to circulate online of the song being performed during a concert by the members of an all-female orchestra from Béjaïa, a coastal city in the predominantly Amazigh region of Kabylia (*Tamurt n Leqbayel* in the Kabyle language, منطقة القبائل in Arabic). The Association Ahbeb Cheikh Sadek el Bejaoui perform traditional Andalusi repertoire and have undertaken tours throughout North Africa and Europe. The event that appears in this recording took place at the *Centre Culturel Algérien* (Algerian Cultural Centre), a venue in the 15th arrondissement of Paris that was founded in 1983 and provides opportunities for Algerian artists and musicians within the French capital. The concert may have been taking place outside of Algeria, but mobile phones were highly visible among the audience members and the ensemble were surely aware that they were being recorded and that these recordings would be shared with audiences in North Africa. As such, the musicians were making a highly visible public statement of support for the *Hirak*, while placing themselves in a degree of danger of public admonishment or worse upon their return to Algeria. The fact that an ensemble that is almost entirely made up of women, within the context of Algeria's patriarchal society, makes their performance of *La Casa del Mouradia* even more significant.

However, the relationship between Andalusi music, the Algerian state, and discourses of nationalism is a complex one. On the surface, Andalusi embodies a socio-economic profile similar to that of those within positions of power, and the music regularly garners financial support and favourable representations within the media. Straddling both the past and present, Andalusi simultaneously engages with notions of the local, regional, and pan-Maghrebi, with Dwight F Reynolds (2020: 7–8) writing that,

> In many places in the modern Arab world, this repertory, while acknowledged to have originated in medieval Muslim Spain, is also embraced as being a vibrant part of local or regional identity. This is a tradition, therefore, that is both medieval and modern, as well as Andalusi and pan-Arab and intensely local, all at the same time.

This produces a specific set of circumstances within which the status of Andalusi is often in a state of flux, and Jonathan Glasser (2016) writes that 'the concept of Andalusi musical patrimony constitutes a special kind of subject: impassive yet powerful, simultaneously national and profoundly other, continually shifting its locus between the centres of power and its margins' (2016: 19). Carl Davila (2015) notes that, despite its apparent position of privilege within Algerian society, the Andalusi nūba has often resisted attempts to align itself with postcolonial nationalism, and he suggests that,

> The Algerian tradition underwent a revival and preservation process in the late nineteenth and early twentieth centuries ... But unlike in Tunisia and Morocco, the state has had less influence in this process, so that the nūba traditions in Algeria have hitherto been less strongly associated with discourses of nation and modernity.
>
> *(Davila, 2015: 163)*

As such, we might understand the performance of *La Casa del Mouradia* by an Algerian Andalusi ensemble as part of a wider challenge to state power, made possible here by the emergence of the *Hirak* movement. The *Hirak* has, I suggest, disrupted the very notions of national and other, of centres of power and margins within Algeria, beginning the process of restructuring society, at least in part, through musical and sonic expression.

Furthermore, much like Torki's appearance on Sky News Arabic, the recordings made on mobile phones at the concert capture a particular sonic event, which is then mediated and disseminated through audio technologies and file sharing platforms. It was the circulation of these recordings, particularly via social media, which made the ensemble's performance both sonically and politically meaningful to Algerian audiences, both in North Africa and among diaspora networks. And it is here, within the ephemeral quasi-public spaces of the Internet and social media, that we find the largest shifts enacted by the songs and sounds of *Hirak*. There is, of course, nothing new in noting that social media has radically reshaped political movements and acts of dissent around the world. However, I suggest that this is perhaps the first time that such communicative technologies have so clearly connected the sonic and musical with the political within an Algerian context. Unlike previous moments of socio-political tension within the country, such as the *Tafsut Imazighen* and *Fitna*, the *Hirak* movement has been able to sonically articulate its message beyond national borders and class boundaries, marrying the audio and visual in the sensorial representation of protest.

Hip Hop as Virtual Political Action

While football stadia provide a physical safe space within the Algerian public sphere, the Internet offers an alternative form of protection from the authorities for those wishing to express political dissent. The Algerian government continues to monitor online activity within the country, but the use of proxy servers and VPNs (Virtual Private Networks), as well as the enduring prevalence of public Internet cafes, means that file sharing and social media have become important means of political debate and expression among Algerians. Furthermore, the long history of Algerian emigration to Europe, and France in particular, has ensured a steady flow of both musics and alternative political opinions both from and back into the country via extensive diaspora networks.

Much like the supporters of the USMA football club, musician Raja Meziane was using her songs to engage in anti-government politics long before the emergence of the *Hirak*. Meziane was born in north-eastern Algerian in 1988 and performed widely as a singer and rapper during her university studies in Law. Following graduation, she developed a number of musical projects, many of which openly criticised the authorities and quickly attracted the attention of the security services. Meziane came to public prominence in Algeria with songs that were critical of the government and *le pouvoir* posted on her YouTube channel in 2013, and soon found herself blacklisted by record labels, radio stations, and television channels, apparently following orders from the authorities. Censorship was not the only concern that she experienced and Meziane was subjected to a number of threats in the lead up the 2014 Presidential elections, recalling that 'I used to receive phone calls from people blackmailing me. "Either you sing in support of Bouteflika's fourth term, or you forget your career in Algeria completely"' (Benaissa, Cassel, and Noman, 2019). Given the history of musicians being targeted with violence in postcolonial Algeria, including a number of murders during the *Fitna* of the 1990s, these threats were not taken lightly, and she emigrated to Prague in 2015.[13] Meziane remembers the trauma that this experience produced, recalling that 'I left my mum. I left the country that I love to death and I never imagined I would leave. But I had to. I had to breath freedom, I had to work' (Benaissa, Cassel, and Noman, 2019).

Life in Prague provided Meziane with a greater degree of creative freedom than had been possible in Algeria and she began to release new music online, which found an enthusiastic audience both in Algeria and among the country's diaspora communities. Her public profile increased when, in March 2019, she released a song entitled '*Allo le système*' (Hello the System), which denounced the Bouteflika regime and Algerian political system, while voicing support for the protests taking place on the streets of Algerian cities. Drawing upon many of the musical tropes of the Trap subgenre of hip hop, the track features highly processed rhythmic elements and a repetitive synthesiser line that clearly represents the sound of a police siren. Autotune is used as a creative device on Meziane's vocals, with the resulting sound producing a lack of clarity, which alludes to the sound of a telephone call, and her lyrics are often difficult to decipher, even for speakers of *Darija*. However, the accompanying music video makes the song's narrative clear, with Meziane depicted emptying her pockets of loose change and her Algerian passport, before making a call to '*le systeme*' (or *le pouvoir*) via a public telephone in a Prague Metro station.

The music video for the song intersperses shots of Meziane performing in the Prague Metro station with clips of protestors on the streets of Algerian cities, simultaneously underscoring her physical separation from the events of the *Hirak* while placing her within the same political discourse as the protestors. Her lyrics are candid and direct, attacking the system for their alleged corruption and the indifference shown towards the Algerian public, and revolve around a chorus in which she acerbically claims that the Algerian people demand a democratic republic rather than a totalitarian monarchy. Given the promises made by the FLN government after national independence in 1962, of creating an egalitarian postcolonial socialist state that would serve the interests of the Algerian people, Meziane's allusion that the country has in fact become a 'monarchy', with a clear line of succession, pinpoints many of the frustrations embodied by the *Hirak*.

We might read the song, and its accompanying video, as a conversation between Meziane and the Algerian political establishment, or a form of remote monodirectional dialogue produced by her effective expulsion from the country. Her anger is undoubtedly directed towards *le pouvoir* and her message explicit, leaving listeners in little doubt of her support for the *Hirak* and its aims. Both song and music video have proven extremely popular and influential among listeners. At the time of writing, the video for *Allo le système* on Meziane's official YouTube channel has 82 million views, with a number of other versions of the song on the same platform garnering further interest and public exposure.[14] This is certainly not an insignificant number of listeners for a political hip hop track performed in a vernacular language spoken in a specific area of North Africa.[15] For Meziane, it is important that the song is recognised as not simply commenting upon Algerian politics and the emerging movement, but rather as a way of engaging from afar with the protests taking place. Almost from the moment of the song's release, Meziane has been clear that she is not simply commenting upon Algerian politics and the *Hirak* movement, but rather that her music enables her to take an active role in the protests taking place across her country. She suggests that the public should recognise that 'I didn't make this song about the popular movement, but to be part of the popular movement. Sadly, we can't go back to our country but we're here. We're with you' (Benaissa, Cassel, and Noman, 2019). As such, the song provides Meziane with a form of agency in that it allows her to speak directly to those still in the country, and somewhat of an active role in the anti-authoritarian politics emerging in Algeria at the time.

Allo le systeme thus provides another example of the ways in which musicians and Algerian citizens have used music and sound over the past four years to disrupt binary notions of physical and virtual space through their activities as part of the *Hirak*. From Sky News cameras and memes to football terraces, concert halls and YouTube videos, the *Hirak* has led to the creation, distribution, and circulation of musics and sounds that cut across, and often collapse, ideas of online and offline engagement and private and public space. Furthermore, they have done this while challenging existing understandings of the hierarchies that structure the social spectrum within Algeria. The collective voice of the *Hirak* has incorporated the oppressed and those on the margins of Algerian society, and those in positions of prestige and privilege, representing the broad range of lived experiences that exist within contemporary Algeria. Returning to the work of Ben Tausig (2018), he employs the term 'sonic vernaculars' to interrogate localised forms of protest, arguing that 'sonic upheaval lies at the heart of how public dissent is experienced and, indeed, of why it is usually staged at all' (2018: 40). Such a term seems particularly apposite for the role of music and sound within the *Hirak*, a movement formed from a uniquely Algerian set of social, political, and economic conditions, both past and present. The 'sonic upheaval' here has been manifested in voices of dissent, acts of musical provocation and critique, and a general shift in the ways that music and sound engage with political discourse and notions of social justice. While the initial aims of the *Hirak* movement are far from being fully realised, radical change has occurred within Algerian society, with those previously situated at the periphery of society suddenly afforded a chance to be heard, and those who might perform on a concert hall stage happy to follow their lead.

Conclusions

On the 17 of September 2021 Abdelaziz Bouteflika died of a cardiac arrest at his house just outside Algiers. He was 84 and had been suffering from ill health for a number of years. The news was undoubtedly momentous within Algeria given Bouteflika's influence on postcolonial Algerian politics, and obituaries appeared within the international media. An article on the BBC website, reporting on the former President's death, suggested that,

> Those who will remember him fondly will recall Bouteflika's revolutionary past in the Algerian war for independence and his time as an outrageously skilled diplomat in a newly independent country. However, many will also point out that he overstayed his time as president and facilitated corruption by surrounding himself with power-hungry oligarchs.
> *(Mezahi, 2021)*

While the article attempted to provide a balanced overview of Bouteflika's life, including acknowledging the motives behind the emergence of the *Hirak*, the report failed to reflect the ongoing tensions and uncertainty within Algerian society, and the fact that for many people in the country, the same issues of social inequality and clandestine power structures remain.

15 months later, on the 14 of December 2022, the members of Ouled el Bahdja announced that they would be disbanding the ultras group after 13 years. A report on the announcement, published on an Algerian football website, argued that,

> At the time of closing this chapter, we can say that the supporters of this group of Ultras is ultimately not specific to USM Algiers. Indeed, for many Algerians, Ouled El-Bahdja was a spokesperson for the people. In particular, with the song "Casa d'El Mouradia" (sic) which became an official song during the *Hirak* in 2019. This is to say that all of this goes beyond a purely footballing framework. Moreover, the reasons that led to this decision result from the repercussions of OEB's socio-political commitment.
>
> *(Touileb, 2022)*

Almost four years on from the initial events of February 2019, the assertion here is clear: that the censorship and threats received by the members of the group, as a direct result of their anti-government political engagement, are ultimately responsible for the demise of Ouled el Bahdja. While their songs helped to shape the *Hirak* protests and provided a soundtrack that went far beyond the football terrace or the urban street, the group found themselves constrained by the very system that the *Hirak* has intended to reform.

Since its beginnings in early 2019, much of the public, media, and political discourse around the *Hirak* movement has drawn upon tropes commonly associated with sound and listening. The country's football supporters are frequently described as the 'true voice' of the movement. Protestors have spoken of making their voices 'heard' for the first time, not only in an abstract political sense but also physically on the streets of the towns and cities in which they live. The members of *le pouvoir* have regularly been accused of not 'hearing' or 'listening' to the country's citizens. When, in December 2019, Abdelmajid Tebboune was finally elected as the new President of Algeria, one media report was quick to argue that 'the message of Hirak has not reached the ears of *le pouvoir*' (Charrier, 2019). While deconstructing *Algerian* political power structures and rebuilding a more equitable society might remain a considerable way off, Algerian citizens continue to find ways of calling for social justice through the *Hirak*. Protestors on Algeria's streets have carried signs proclaiming, 'my voice is sacred, like my dignity and my liberty.' Perhaps, if Algerians can continue to find new and innovative ways of expressing their opinions and demands through sonic dissent, their desires might eventually be met and Sofiane Bakir Torki's call for real change might finally come to pass.

Notes

1. Bouteflika was born in Morocco in 1937, the son of Algerian parents. After joining the pro-independence FLN, he quickly rose through the ranks, becoming an important political figure after independence in 1962. Having served four terms as President of Algeria, he died at his home outside Algiers in September 2021.
2. Fitna (فتنة) is an Arabic term that translates loosely as 'struggle' or 'trial.' Historians of Algeria have been keen to avoid using the term 'civil war' for the conflict that took place throughout the 1990s.
3. Pouvoir is a French word that translates into English as power. Within this context, it is useful to understand the term as analogous to 'the powers that be.'
4. Hirak (حراك) is an Arabic term that means the 'movement.'
5. The National Liberation Front (*Front de libération nationale* in French, or جبهة التحرير الوطني in Arabic) began life as an anti-colonial political movement and assumed power after national liberation from French rule. They have remained in power ever since.
6. Torki used the term 'Ytnahaw ga', an expression in Darija which can loosely be translated as 'remove them all' or 'remove them altogether.' For Algerian viewers, there was little doubt that Torki was referring to *le pouvoir* and the country's entire political system.
7. Fusha (العربية الفصحى), often referred to as Modern Standard Arabic (MSA) emerged in the late 19th century as an attempt to standardise and regulate both written and spoken Arabic. While the language is used in official contexts and often taught in schools throughout the Middle East and parts of North Africa, it remains associated with the educated middle classes, with local vernacular forms of Arabic differing between region and country.
8. al-Hogra (حقرة) is a term in *Darija* that denotes a sense frustration or anger at injustice.
9. Evans is referring here to the ways in which the government has employed speculative statistics in commemorating the deaths of Algerian citizens during the period of French colonial rule. The exact number of Algerian casualties remains unknown, and elsewhere Evans notes the adoption of what he terms the 'war of one and half million martyrs' (2022).
10. Algeria has a long history of religious diversity. However, following independence in 1962, many members of the country's Jewish communities migrated, in particular to France and Israel.
11. The 'Sporting Union of the City of Algiers'.
12. Ultras is a term used globally for the most committed and passionate supporters of football clubs. Ultras are often associated with singing and chanting, and visual displays using flags and banners (known as tifos). Many ultras groups have strong associations with political ideologies and movements.
13. A number of well-known musicians were murdered during the 1990s, including Matoub Lounès (discussed previously), as well as the *raï* singer Cheb Hasni

and produced Rachid Baba-Ahmed. As a result, many commercially-successful musicians (and *raï* singers in particular) left the country, often moving to France.
14 Correct on the 16 of January 2023. Additional videos on YouTube that feature *Allo le Système* include a number of 'reaction' videos, in which Algerian and non-Algerian YouTube users respond to the original song and music video, often explaining the concept of the *Hirak* to non-informed viewers.
15 To provide further context to these figures, the population of Algeria in 2021 was under 45 million citizens.

References

Amara, M. (2012) Football Sub-Culture and Youth Politics in Algeria. *Mediterranean Politics*, 17(1): 41–58.
Bella, N. (2019) Shouting for a New Algeria: Slogans as Foundations of a Political Project?. *Arab Reform Initiative*, 11 Dec, available at: www.arab-reform.net/publication/shouting-for-a-new-algeria-slogans-as-foundations-of-a-political-project/. Accessed: 2 July 2023.
Benaissa, F., Cassel, M., Noman, M. (2019) Raja Meziane: Algerian Unsigned Singer becomes Sound of Revolution. *BBC*, 24 Oct, available at: www.bbc.co.uk/news/av/world-africa-50160646 Accessed 1 July 2023.
Bentahar, Z. (2021) "Ytnahaw Ga!": Algeria's Cultural Revolution and the Role of Language in the Early Stages of the Spring 2019 Hirak. *Journal of African Cultural Studies*, 33(4): 471–488.
Benrabah, M. (2013) *Language Conflict in Algeria: From Colonialism to Post-Independence*. Berlin: De Gruyter.
Berkani, M. (2019) Algérie: "Un Million de Personnes Dans La Rue et Aucune Image à La Télévision". *Franceinfo*, 25 Feb, available at: www.francetvinfo.fr/monde/afrique/algerie/election-presidentielle-en-algerie/un-million-de-personnes-dans-la-rue-contre-le-5e-mandat-de-bouteflika-et-aucune-image-a-la-television-les-algeriens-indignes_3206611.html. Accessed: 2 July 2023.
Boundaoui, A. (2015) Algerian Football Fans Sing Songs of Freedom. *Middle East Eye*, 13 Feb, available at: www.middleeasteye.net/features/algerian-football-fans-sing-songs-freedom. Accessed: 30 June 2023.
Charrier, A. (2019) Algérie: "Le message du 'Hirak' n'est Pas Arrivé aux Oreilles du Pouvoir". *France24*, 19 Dec, available at: www.france24.com/fr/20191219-alg%C3%A9rie-le-message-du-hirak-n-est-pas-arriv%C3%A9-aux-oreilles-du-pouvoir. Accessed: 4 July 2023.
Correia, M. (2019) The Soccer Fans that Toppled a Government. *The Nation/Le Monde Diplomatique*, 15 May, available at: www.thenation.com/article/archive/algeria-bouteflika-fans-protests/. Accessed: 30 June 2023.
Davila, C. (2015) The Andalusi Turn: The Nûba in Mediterranean History. *Mediterranean Studies*, 23(2): 149–169.
Evans, M. (2012) *Algeria: France's Undeclared War*. Oxford: Oxford University Press.
Evans, M. (2022) Is Algeria Still Defined by it Liberation Struggle?. *History Today*, 7 July, available at: www.historytoday.com/archive/head-head/algeria-still-defined-its-liberation-struggle. Accessed: 28 May 2023.
Glasser, J. (2016) *The Lost Paradise: Andalusi Music in Urban North Africa*. Chicago: University of Chicago Press.

Jarvis, J. (2021) *Decolonizing Memory: Algeria and the Politics of Testimony*. Durham, NC.: Duke University Press.

Lowi, M. R. (2010) *Oil Wealth and the Politics of Poverty: Algeria Compared*. Cambridge: Cambridge University Press.

Marouf, N. (2002) Le Système Musical de la San'a ou le Paradigme de la Norme et de la Marge (Hommage à Pierre Bourdieu). *Horizons Maghrébins*, 47: 8–24.

Meddi, A. (2019) Manifestations du 22 Février: Pourquoi les Algériens sont en Colère. *Le Point Afrique*, 24 Feb, available at: www.lepoint.fr/afrique/manifestations-du-22-fevrier-pourquoi-les-algeriens-sont-en-colere-24-02-2019-2295841_3826.php. Accessed: 3 July 2023.

Mezahi, M. (2021) Abdelaziz Bouteflika, Algeria's Longest-serving President Dies. *BBC*, 18 Sep, available at: www.bbc.co.uk/news/world-africa-56269634. Accessed: 4 July 2023.

Muradova, L. (2016) Oil Wealth and Authoritarianism: Algeria in the Arab Spring. *Revista Española de Ciencia Política*, 40: 63–89.

Reynolds, D. F. (2020) *The Musical Heritage of Al-Andalus*. London: Routledge.

Schaefer, C. (2015) Who Killed Matoub Lounès?. *World Literature Today*, 89(6): 29–31.

Shannon, J. H. (2015) *Performing Al-Andalus: Music and Nostalgia Across the Mediterranean*. Bloomington: Indiana University Press.

Tausig, B. (2018) Sound and Movement: Vernaculars of Sonic Dissent. *Social Text*, 136(36/3): 25–45.

Touileb, M. (2022) USM Alger: Ouled El Bahdja, le Chant de Cygne. *La Gazette du Fennec*, 16 Dec, available at: https://lagazettedufennec.com/usm-alger-ouled-el-bahdja-le-chant-de-cygne/. Accessed: 4 July 2023.

7
BORDER SPACES AND SOUNDS OF RESISTANCE

Music at the Franco–British Border

Celeste Cantor-Stephens

Introspection/Outrospection

A Little Context

The ocean sloshes and growls under the hull of our boat, occasionally slapping up against its sides, as we compliantly rise and dip. Deep rumbles of gently rolling waves are a constant reminder that the sea is holding us up. A familiar drone of British voices wanders the decks, with family holiday plans seeping dryly between the chatter of what feels like a million French schoolchildren heading home from a class trip away.

 The journey, on the ferry, only takes 90 minutes. Barely any time passes between the safety announcements at Dover, and the tannoy call to declare our arrival in Calais. When we reach the port and file off the boat—by bicycle, in my case[1]—I stream down the ferry ramp and through the maze of port pathways before reaching the roads that eventually lead into town. Over the years—between the first time I completed this journey at the beginning of 2015, and today, in 2023—the experience changes. Most notably, there are more gates to go through, and endless, high metal fences have been erected, stretching along the roads alongside the port, and upwards, towards the sky. They are topped with razor wire, and even though I, with a UK passport in my hand, am able to walk around them, the looming metal, and sharp points feel menacing and uncomfortable. Occasional shreds of clothing, snagged on the barbed wire and flapping frantically in the wind, emphasise my worries and prompt questions about the events that came before.

These were the kind of questions and events that first brought me to such border spaces, as an activist, in support and solidarity with the people most affected by these fences and restrictions. Seeking deeper ways to understand and share stories, I incorporated research, for master's degrees, and subsequent independent projects, in written and sound-based forms. Still prioritising active solidarity, my role as a musician draws my attention to the importance and meaning of sound in these spaces. As I cycle away from the ferry port, the sloshing of the ocean, clinking boat masts, and whirring harbour turn into the sound of lorries passing on the nearby motorway, and the police vehicles that drive by to wait, conspicuously, at corners.

Borders are complex, although we may often be encouraged to believe in their simplicity. At their most basic, we might imagine them as lines on a map, as points to pass through, as moments of governance where rules are applied as people enter or exit a country and step into a new space, with its own rights and rules. I argue that borders are malleable constructions of the state and of those in positions of power. It is precisely because borders can appear absolute, because they are in Balibar's (2004: 109, original emphasis) terms, '*absolutely nondemocratic*,' that they are also, in practice, ambiguous, flexible, and manipulable; subject to change and the whims and prejudices of *certain* people. Events at the Franco–British border demonstrate this. With a focus on the experiences of displaced people in Calais, Northern France, this chapter explores these ideas. Here, I propose that borders are spaces of exception, where rights are replaced with brutality. Rather than being clearly delineated lines, borders can spread outwards, move *within* a territory, and, for some, even become ubiquitous, inescapable. I posit a concept of a *Certain Human's Land*, identifying the border as a space governed by some, to the detriment of Others. These borders are, ultimately, a production, and a tool of structured inequality.

Despite these complex and punishing settings, music persists. Although it may not always be a priority or practical, music has such a pronounced value that many at the Franco–British border go to great lengths to access it. Centring this chapter around a selection of musical moments I encountered in Calais between 2015 and 2023, I highlight the role of music in this context. From supporting momentary escape, to reclaiming space, consolidating community and individual ownership of experiences, assuming agency, or making explicit political statements, music functions as a source and support of resilience and resistance. Although not always explicit, music is a meaningful, relatively accessible, and, for some, crucial tool in resisting and fighting inequality and injustice at borders. A potential affirmation of humanness in an often-dehumanising context, music encourages connection that transgresses borders, to support wider recognition of their constructed, destructive, and unjust nature.

The stories I include here stretch far beyond the fragments depicted, and my understandings (or possible *mis*understandings)[2] are inevitably and unavoidably influenced by my own position and experiences. Other people have contributed enormously to this chapter, sharing their experiences and allowing me into their lives, often under exceptionally challenging circumstances. I am very grateful. All those whose names, words, and stories I include here have given me permission to do so, allowing their creative ownership to be recognised. For others, identification carries risks such as persecution, asylum rejections, or distress for separated families; some contributors remain anonymous.

This chapter takes an unconventional form, combining three different voices: the discursive and analytical writing you might expect from a typical academic piece; stories, vignettes of real encounters and experiences, formed from scraps of diary entries, memories, and other documentation, in *italics* for clarity; and, finally, a Piece to conclude each section. I hope that the inclusion of human stories—and readers' human responses to them—will support understanding of the events and surrounding discussion. The Pieces take the form of text-based scores,[3] to perform alone or with others, to stimulate further thought and feelings, and to support readers' explorations of their own experiences and place in (and beyond) the context of this chapter. These original Pieces are influenced by practices in Deep Listening and the work of composer Pauline Oliveros.[4] They are practical actions which can bring insight and help to facilitate a shift in awareness. They are not essential to understanding the main body of the text, and if you find them excessively challenging or disturbing, you are free to pass them over, or to read them as poetic intervals.

PIECE
Centring Outwards/Your Borders[5]

Imagine yourself as a point in the centre of a circle.

Sitting, standing, or lying down, take a few minutes to listen to the sounds that come from you, that are part of you, that are made by you.

After a few minutes, spread your attention outwards. Listen to the wider sphere of sounds in your surroundings.

Spread your attention further still, to include sounds beyond the borders of your current listening, and the borders of your imagined circle. What can you hear if you really focus outwards, further than your usual stretch of attention?

Repeat these stages. Begin by returning to listening to yourself, at the centre of your circle.

A Swirling Chaos of Exception

Space(s of Exception)

From across a sea of tents and ramshackle shelters, I could hear a distant call of deep percussion. Winding down little passages of dirt-ridden scrub, stepping carefully over uneven ground, old rubble, and between tents placed impossibly close together, more came into range. Sounds of rubab[6] strings and an electronic synthesizer, still separated by enough tarpaulin to sound like they were underwater. At last, I stepped onto the edge of a small clearing between tents and met the source.

In 2015, on the outskirts of Calais town, near motorways, factories, and away from the homes of long-term locals, a sprawling mass of tents spread tightly across sandy land, strewn with ancient rubbish. Adults and children from Eritrea, Afghanistan, Iraq, Syria, Pakistan, Sudan, Ethiopia, Iran, and many more countries, inhabited this makeshift refugee camp on the northern French coast. Some aimed to cross the 21-mile stretch of sea to the UK, others sought asylum in France, and some hoped for refuge in different countries. The space was defined by unpredictability and vulnerability. At its largest, somewhere around ten thousand[7] people (Refugee Rights Europe, 2018) were crammed onto the land, where local authorities would, for that moment, tolerate their presence (Seba, 2015; Schlembach, 2016). Despite these conditions there was no real sense of safety or security, no protection for the vulnerable except that given by the community, volunteers, and visiting charities, and, for a period, a limited-capacity shelter for women and children. Anti-migrant groups were known to attack displaced people near the site, CRS[8] police would throw in teargas grenades from the overlooking motorway, and fires would spread rapidly from tent to tent, with few routes through for emergency services.[9]

Despite its volatility, there was some regularity to the shanty-town space. Through the squalor and distress, certain sights, sounds, and interactions became familiar in the space. Warm greetings, invitations to chat and drink tea, fraught questions about the asylum processes and curiosity about the UK, requests for help, wary glances, and semi-joking pleas to travel back with me to England were common. Amongst other regular non-residents were Nadine Rubanbleu and Dominique Mégard. Following requests, the retired Calaisian couple began taking a generator and two large speakers into the camp, where communities would take turns to play music via mobile phones. Sometimes a short dance party would erupt, with people spinning around in brief moments of release.

Drawn across the camp by the sounds from the speaker-generator system, I found myself at the clearing between tents. As traditional Afghan instruments melded with more contemporary pop arrangements and the occasional hit of lurid European dance music, I was struck by the scene. This wasn't a dance party or a moment of intentional, collective listening. Nor was it that familiar scene: the warm welcomes, the questions, the clear, exhausted desperation that comes with living among rubble and dirt and tents that blow down in coastal night winds. While none of this had disappeared, the music-centric space seemed abnormally ... normal. Enveloped in sound, a large group of people had gathered. A few crouched near the speakers, casually playing DJ as they cued up music; many sat on the ground, chatting, and charging phones at the generator. Some were playing a game of net-less volleyball, their onlookers apparently unfazed as the ball narrowly missed them and whizzed past tents. Largely, people stood, in a space held by music, talking, watching, being.

The music in this space was big and obvious and dominant, but it was also, in many ways, peripheral, *not* an explicit, collective focal point. It would be naïve and dismissive to suggest that the heavy context surrounding this bubble-like moment had been forgotten by anyone within it. But, weaving between people, through activities, and the area around them, sound acted as if a binding force, underpinning a space that appeared, momentarily, separated from the surrounding, swirling chaos of abuse, vulnerability, and exception.

In a *state of exception*, laws and norms can be suspended, violated, by the sovereign power (Agamben, 2005; Schmitt, 2005). This concept, and its legitimisation, is often associated with jurist and theorist Carl Schmitt (2005), a critic of parliamentary democracy, who was, for a period in the 1930s, a prominent member of the Nazi Party (Ohana, 2019). Despite this, Schmitt's concepts are still visible in today's governing paradigms (Agamben, 2005). Based on the idea that in exceptional circumstances the state may take exceptional measures, the concept of the state of exception is founded in paradox; these measures and anomies fall outside of the law, yet are inscribed from within the juridical order, applied by the powers who, ostensibly, oversee and enforce legal systems. As philosopher Giorgio Agamben (2005: 1) outlines, 'the state of exception appears as the legal form of what cannot have legal form.' This state of exception is enacted at the border, a *space* of exception where conventions and rights are overwritten by those who are meant to uphold them. In Calais, from 2015, while other makeshift sites were demolished, a single camp was tolerated by the municipality, in an area contaminated with dangerous industrial substances (Hicks & Mallet, 2019). At the time of the sound system[10] gathering referenced above, there was no running water, electricity, or lighting on the land: unfit for human living, despite being recognised by authorities. It was frequently attacked with tear gas by the CRS police, and stories of further police brutality were commonplace (Calais Migrant Solidarity, 2015a, 2015b).

In October 2016, the slum-like camp was demolished by authorities, making those who would have lived there even more vulnerable. Charity-donated tents popped up in areas hidden away from the town, but they lacked the eyes and solidarity of surrounding community. Recurring accounts from other people, along with events I witnessed first-hand, highlight behaviour patterns of the police since this time. These include patrolling public parks with pepper-spray in hand, arresting people queuing for food distributions, waking people up in the depth of night using truncheons and lachrymatory tear gas,[11] spraying gases directly into the eyes of adults and children, covering blankets with pepper-spray to make them unusable, stealing sleeping bags, beating people with batons, seizing people and abandoning them miles away from Calais centre, stealing shoes, destroying phones, throwing possessions into the river. '[The police] are always coming, and they beat people, [they beat] so many people here ... they have broken legs, broken hands' someone told me, in early 2017.

Though *exceptional* in their brutality, these acts appeared ubiquitous. For every volunteer or displaced person I spoke to, intimidating, often dangerous police interactions seemed to be an integral part of Calais existence. One person explained:

> Sometime[s] we [sleep for] one, two hours somewhere, [then police come] and [spray] gas. We run away. And there we try to find other place, and there, when we sleep [...] some other police appear.

Another told me:

> If we tell [the police] we are peaceful people and we don't want [them to use pepper spray], they [spray] our eyes ... they have power, they can do whatever they want. And [they even] told us: "We have power, we can do whatever we want."

But the measures at this *border space of exception* stretch far beyond the police. Led by Mayor Natacha Bouchart since 2008, the Calais municipality has, for example, banned charities from distributing water and food on multiple occasions (Go, 2017; La Cimade, 2022). Meanwhile in the UK, the Hostile Environment policy aims to make life unbearably difficult for displaced people, affecting fundamentals such as healthcare, finances, and housing (Kirkup and Winnett, 2012). Originally introduced in 2012, the government has extended the policy and its impact over time (Travis, 2017). In 2018, the public unfolding of the Windrush scandal[12] forced the UK government to acknowledge some of the damage inflicted by these policies, and suspend certain measures (Rudd, 2018; Grierson, 2018). Despite the acknowledged injustice, primarily affecting displaced people and ethnic minorities, 2023 has so far seen a reintroduction of such anti-immigrant policies (Gentleman, 2023). Just two examples of policies that intentionally disrupt the everyday lives of those at the border, the Hostile Environment and Calais distribution bans have been deemed unlawful, including by the Equality and Human Rights Commission (2020) and Le Tribunal administratif de Lille (2022),[13] respectively (Gentleman, 2020).

The police behaviour described above fits readily into this picture of exceptional behaviours. Supposedly representing authority, the police are physical agents of the political world. They deal directly with the bodies that compose public and non-public spaces: monitoring, arresting, moving, detaining. At the Calais border, they are funded by French *and* British states. In November 2022, the UK government announced it was contributing up to £62.2 million to Franco–British border controls, the latest in a combined sum of over £300 million since 2014 (Gower, 2022). This includes investment in surveillance technology, detection dogs, French 'reception centres,' 'removal centres,' and 'the deployment of more French officers' (Home Office, 2022; Gower, 2022). Here, British funds contribute to the creation and preservation of these systems of exception, extending beyond the borderline, and even beyond the UK, into France.

And so, six years after the notorious Calais camp was dismantled, people continue attempts to establish small, temporary spaces of shelter, and to avoid the hard hands of the police who regularly tear them down. While this is happening, even without a stable base, new sound system moments break through. The organisation *Mosaïc* was established in 2021 to support exiled people in Calais to access arts and culture. Today, they bring the speakers to carefully located, out-of-town sites, away from the immediate threat of police, so that people might have some relief from this space of exception, in the form of dance and sound.

Julian Henriques (2003) uses the term 'sonic dominance' to describe an experience that is exemplified in the reggae sound system. Here, sound is not just heard, but felt over the body. It is a visceral, immediate, and unmediated saturation of sound, inside of which people become lost. 'Hard' and 'extreme,' the sounds of sonic dominance are 'also soft and embracing,' creating a space where 'sound allows us to block out rational processes' (Henriques, 2003: 451–452). Though not a reggae dance party, there are echoes of these ideas in the sound system scene of the camp, where music seemed simultaneously hyper-present and almost incidental.

Sonic dominance isn't reserved for the dance hall. It exists in other contexts, including, as Henriques notes, political demonstrations. There are parallels between protest and reggae sound system culture, rooted in resistance to oppression and inequality (Prince Buster in Bradley, 2001: XV). Like the Calais version, the reggae sound system emerged as a relatively accessible instrument among dispossessed communities, and as a means of creating space controlled by otherwise-subjugated voices, without interference by authorities (Veal, 2007; King, 2002). While it might be less explicit than marching among a crowd of protesters crying 'Whose streets? Our streets!' the sonic take-over of a space can be accompanied by a sense of ownership, taking charge and assuming one's own value and identity; sound is 'both a source and expression of [power]' (Henriques, 2003: 453; Hebdige, 1979). When a space becomes dominated by an authority, as and transformed into a place of *exception* and *exceptional* abuse, reclaiming it is a powerful event. Small musical gatherings—facilitating scenes that might otherwise appear 'normal'—become another kind of *exceptional space* within the *space of exception*.

PIECE
Variations on Variations[14]

Listen for silence. What else can you hear?
Respond to a sound with an echo or your own sound.

Which sounds are human; which sounds are not?
Respond to a new sound with an echo or your own sound.

Which sounds do you feel part of?
Respond to a new sound with an echo or your own sound.

Which sounds do you feel excluded from?
Respond to a new sound with an echo or your own sound.

Perform this piece in different spaces. Ask yourself the same questions.

Whose Streets?

The Border Expanse: A Certain Human's Land

Calais is windy. Exceptionally so. If the police don't get to them first, it's not uncommon for tents to collapse in the fierce, cold gales that come in from the North Sea or funnel up through the English Channel. On one such wet and blustery day, back in the time of the big shanty-town camp, I was beckoned over to shelter in a small shack, as I tried to work my way across the site, against the wind. Inside, six Afghan men huddled around a little wood-burning fire. Perched on top, tea boiled in a handleless pan, poured into plastic cups, and shared around for warmth. We chatted, sharing stories, and it turned out that Abdullah, crouched on the ground to my left, was a poet, a singer. He pulled some scraps of paper from his coat pocket, carefully unfolding them to reveal hand-written Pashto verses. Backed by the sounds of harsh winds against a shelter of reclaimed wood and tarpaulin, Abdullah's voice gently, delicately took over the space. Through three a cappella songs, he told stories of his recent life.

This is a translated extract from one of those songs:[15]

*Whatever happens is on us
Because we left Afghanistan and came here*

*Some of us are dying on Iran's border
Some of us are dying on Turkey's high mountains
Some of us are dying in the rivers of Greece
How worthlessly we are dying in travel
Now we've left Afghanistan*

*Whatever happens is on us
Because we left Afghanistan and came here*

*Some of us in tents, some of us on the streets
In churches, in parks
Oh my God take pity
Oh my God, in the Jungle*[16]

In the five years it had taken Abdullah to reach Calais from Afghanistan, he had written around 800 songs. And while life at the Franco–British border was clearly traumatic, it was just one part in a continuing story of displacement, trauma, and injustice.

We often see borders as lines: thin marks on a map, points to step over or pass through as we move between countries. But the conditions thrust upon the people at borders stretch far beyond walls, international terminals, and passport checkpoints. The apparently *exceptional* measures associated with the Calais border, including destroying shelters and banning food distributions, extend beyond the border into the town and its outskirts. The Hostile Environment, meanwhile, spreads across a whole country, albeit to a targeted population.

Detention centres, often set away from international frontiers, are further examples of the expanding border. Termed 'Immigration Removal Centres' by the UK government, these prison-like institutions are, in theory,[17] used to hold non-residents for administrative purposes, before they are deported from the country or as they 'wait for permission to enter the UK' (McGuinness and Gower, 2018: 3). These border spaces, these *between*-spaces, where individuals are held without a time limit, are spread across the country, extending far from the borderlines we see on a map (Home Office, 2023; UK Government, n.d.). Somewhat paradoxically, the UK also runs multiple 'short-term holding facilities' in Northern France; that is, on French land but under British operation. These extra-territorial controls, outsourced to private, profit-making companies, Mitie and Eamus Cork (HM Chief Inspector of Prisons, 2020), are flagged by The Detention Forum (2022) as 'legal and procedural grey zones.' As if not-quite-in-any-country, ambiguities mean that the UK government 'is able to shift or indeed entirely deny responsibility for those held in the centres' (The Detention Forum, 2022). Correspondingly, a report by HM Chief Inspector of Prisons (2020) found 'serious concerns' that the 'Border Force could not tell us the legal authority under which [certain detainees] were held.' Deportations and related systems of enforced movement can also behave like extended, rule-avoiding border spaces. The Dublin Regulation, or Dublin III,[18] allows Member States[19] to send people back to a country they have already passed through, refusing responsibilities, and potentially thrusting refuge-seekers into perpetual statelessness and abuse (ECHR, 2021). Like British-run 'holding facilities' in France, the Dublin procedure creates a self-contradicting (non-)space of simultaneous detention and movement: a kind of passing through the border that never seems to end.

Of course, the shift from border*line* to extended border *space* is typically reserved for a specific group of people. This is exemplified in the French *zone d'attente* ('waiting zone'), a holding space for 'foreigners' being denied entry into France. While many *zones d'attente* are at international ports, their exact number and location is difficult to determine, since they can be temporarily created and deactivated as authorities see fit (Makaremi, 2009; Ministère de l'intérieur, 2022). The first *zone d'attente pour personnes en instance* ('waiting zone for people in proceedings'), at Roissy-Charles de Gaulle airport in Paris, was in a hotel. On the ground floor, paying guests comfortably enjoyed the rights and norms of Being in France. But on the floor above them, around 25 rooms had been declared a waiting zone, an international space (Makaremi, 2009; Gisti, 1991). For some, Hotel Arcade was, of course, firmly in France; for Others, it was an extension of the border, outside of France, and outside the accompanying rules and rights. Rather than a well-defined line on a map, the border seems to follow specific people. Inescapable, it is as if, no matter where they stand, these people live on the border, even *become* the border. And as they do, the rules shift and mutate around them.

It's tempting to imagine these so-called international zones, these extended borders, and spaces of exception, as a kind of no man's land, or, for the sake of degendering, a *no human's land*. These are spaces that sit between other, better-defined spaces, that seem to exist outside of any clear system of regulations and rights. But space is a resource, an instrument, a means of production, control, and power (Lefebvre, 1991). Rather than a *no* human's land, these border spaces are governed by *certain* people. The 'power-geometry' of space, to use Massey's terminology (1993), is built into the everyday systems that privilege some above others and applies beyond the apparently intentional manipulation of space and power we see at borders. Here, the liberties of some lead to the restricted movement, and effective imprisonment, of others. A complex, self-perpetuating system, those with access to extensive resources draw from those without; with internationally-investing jet-setters at one end of the spectrum (undertaking global travel, consuming resources with environmental consequences across the world), and those confined to inner-city slums at the other. On a more local level, the regular car user (probably unintentionally) reduces the social rationale and financial viability of public transport, increasing their own mobility while potentially reducing the movement of those who rely on that system (Massey, 1993). Certain groups seem to rule over everyone's conditions of mobility, but not always by preventing movement. While displaced people are, or will have been mobile, their movement is typically directed by others: from deporting to detaining, from granting entry to blocking passage, from contributing to environmental crises to providing weapons that force people to move. The system drives itself; as people are forced into perpetual movement, stasis, or both, they lose any control over the space around them, and with it, political rights. In Rancière's (2001) words, '[the] principal function of politics is the configuration of its proper space.' At the Franco-British border, and at borders more widely, shelters are destroyed, communities are split up, and displaced people are unable to gather in significant numbers. These communities are effectively banished, in sight and in sound, from (so-called) public areas, and deprived of any agency or influence over this space. At border spaces, where one party undeniably dominates, the other is denied any opportunity to challenge the hierarchy or to contribute to the transformation of the space and its rules. Those who are denied political input and agency within such a space are thus also denied what should be, in Arendtian terms, the fundamental human 'right to have rights' (Arendt, 1962: 298), and subsequently, the recognition that they are indeed human. Despite being a place where rights, rules and humanness are stripped away, border spaces, in Calais and elsewhere, are not a no human's land; they are a *Certain Human's Land*. Within these spaces of exception, not all rules and not everyone's rights are quashed. Rather, these spaces function according to *certain* conditions, determined by Certain People, to the detriment of Others.

The police in Calais make this clear. As does their funding from the French and British states. When officers are deployed in the name of border security, they are placed in direct opposition to the people against whom the border is being secured. That is, the police have no duty of protection towards displaced people; rather, they are working against them, as supposed threats to border security. In a *Certain Human's Land*, the Other, the displaced, becomes the threat. If a state of exception is a response to the disruption of (purported) normality, or, from another perspective, if legal norms and rights require a homogeneous point of reference, then the presence of foreign bodies is a facile excuse for validating exceptional behaviours. That is, as the border expands to encompass whole areas, those who don't fit into the *Certain Human's Land* become the justification for, and targets of this exception. As Agamben (2005: 2) warns, the state of exception can be used to attack 'entire categories of citizens who for some reason cannot be integrated into the political system.' Turning these entire categories into threatening non-humans is one way of ensuring that they won't be integrated.

'Sometime[s] [police] say: "this is not your country, this is France, you can't stay here,"' a man from Afghanistan explained, as we spoke on the outskirts of Calais in 2017. 'They're thinking "this is my country, you are like refugees, you are nothing."' Across conversations, people echoed similar feelings. One told me: 'I don't think people lived like this 500 years ago.' And another: 'We are human, we need some rights … there are no rights for refugees.' These deprecating and dehumanising experiences, living conditions, and treatments from others are horrific, but they're unsurprising when we consider the similarly dehumanising rhetoric from influential authority figures who wield power over space:

> You have got a *swarm* of people coming across the Mediterranean [...] wanting to come to Britain.
> *David Camreron (BBC, 2015, my emphasis)*
> *then UK Prime Minister*

> [Democrats] want illegal immigrants, no matter how bad they may be, to pour into and *infest* our Country.
> *Donald Trump (2018, my emphasis),*
> *then President of the United States*

> [There is an] *invasion* on our southern coast.
> *Suella Braverman, Home Secretary of the United Kingdom*
> *(Hansard, 2022, my emphasis)*

Back at the Franco–British border, where the rain had eased just slightly, Abdullah and I said goodbye. It's never clear where, when or if those who meet in these border camps will see each other again. But, Abdullah told me, 'If you remember my poems, you remember me.'

Brandon LaBelle (2018: 4) writes of 'sonic agency,' conceptualising sound 'as a set of support structures by which one garners capacities for acting in and amongst the world.' Tia DeNora (2000: 73) describes music as a 'technology of self.' That is, 'as a device for ordering the self as an agent, and as an object known and accountable to oneself and others.' Music can contribute to the creation and affirmation of a person's identity and existence: their social agency, personal and worldly roles, their power, and validity as a human. In dehumanising border spaces, where people are stripped of humanness and individuality, and their agency and rights are taken with them, this affirmation of self becomes even more powerful, as a potential means of strength and resistance in the face of exceptionally undermining circumstances. At one level, Abdullah's pieces show music as a means of exploring and documenting experiences. His songs are a rich and valuable record of events that we otherwise tend only to see depicted, and misrepresented, by the media and authorities. But Abdullah is reclaiming his stories, telling them in *his* words, reappropriating the events and experiences that have largely been dictated by other people.[20]

Crouching under the shelter next to Abdullah was Ibrahim, who had become the poet's closest friend over their time together. In a song bound with humour, Abdullah hyperbolised Ibrahim's role, gently mocking other members of their group with light-hearted jibes. As Abdullah sang, the small community snorted and giggled at each passing caricature.

The whole Jungle[21] seems to me empty
And I miss Ibrahim Jan[22]

Lots of people from this Jungle
Are going through to Lidl[23]
A huge caravan of Afghans
The smugglers fee you can't break with a stick[24]
And I miss Ibrahim Jan

The whole Jungle seems to me empty
And I miss Ibrahim Jan

Look at Qasim, he's getting old
People are laughing at him
He sank and was destroyed like me
look at Wahid, he's getting fat day by day
And I miss Ibrahim Jan

The whole Jungle seems to me empty
And I miss Ibrahim Jan

There is a smuggler in the Shugufans[25]
He's a principled person

Omid is strong and brave
Forget the police; he's even afraid of the security[26]

And I miss Ibrahim Jan
The whole Jungle seems to me empty
And I miss Ibrahim Jan

Music is a means of communicating and exploring events and emotions. Through music 'one may reveal one's self *to* one's self, as well as to others,' write Schneck and Berger (2006: 248, original italics); '[music] is all about just being—human.' Through the humour, Abdullah describes some of the darkness of Calais life. In doing so, he also confronts it, as a human living through it. The reality of these scenes is excruciating, but here, moulded into a song that even manages to be comedic, the events belong to the community, and to Abdullah. In this moment, the story comes across as manageable, even as its protagonists huddle amid a bleak expanse of tents, with coastal weather beating down, and the threat of the border all around.

In the summer of 2022, I sat in a small back garden with Tekle.[27] *He'd been given a bed in the house of a French man, where a few displaced people were able to live at any one time. As we drank tea beneath the sun, Tekle disappeared briefly inside, and returned with a hand-made lyre. Through very limited resources, one of his housemates had managed to find the tools and materials to build an Eritrean krar. He had nailed four small planks of wood into a trapezoid, covered, at the narrower end, with two larger, flatter pieces to form a wooden body. Five strings ran from the body to the opposite frame, held in place by metal screws. It was a beautifully made instrument. From the krar, our conversation turned to Tekle's journey; like the krar-maker, he had travelled to Calais from Eritrea. On his passage across the Sahara, Tekle spent 15 days in a vehicle driven by profiteering people smugglers. 'If we die, they don't care,' he told me. At least one person fell from the vehicle, but the driver didn't stop. This is 'normal,' Tekle explained, '[the smugglers] don't care about the people [on board].' 'It's normal,' he reiterated, but even so 'it's traumatising.' Our discussion moved to the situation around us, and to the police. When volunteers from local organisations are nearby, phone cameras at the ready, the police 'compromise,' Tekle said. But when the people who weren't obviously foreign were gone, 'the police start the violence.'*

A familiar thread in Calais, Tekle's ongoing journey emphasised abuse and indifference from those in control, and a lack of control among the people at the centre of the story. We drifted back to the krar; a topic that felt oddly light in the context of our conversations. I had a go at playing, but wasn't very good. Tekle sounded better, but, he told me, his krar-building housemate was really good. For him it's 'a passion,' Tekle said, 'it's like a dream'.

In this *Certain Human's Land*, some people dominate over others. Tekle's words point to a space (extending not only beyond borderlines, but from the Sahara to Calais) where the apparent validity of one group meets the dehumanising abuse of another. Consistent with ideas drawn from Massey (1993) above, the apparent ownership of space by this first demographic is developed in direct contrast to the deprivation of ownership, and control, from the second demographic. Even the volunteers have more power over events than the displaced people themselves. While certain people gain control, Others seem to be reduced to subhuman. Making music, though, is very human, and an activity that exists across every culture on this planet (Schneck and Berger, 2006). In the dehumanising context of the border space, the 'dreams and passions' in Tekle's words, of making music are an affirmation (intentional or not) of human identity (DeNora, 2000). Finding the means to build an instrument expands on this, an act of agency, and a potentially empowering way of regaining some sense of control. 'Whatever happens is on us / Because we left Afghanistan.' Through these hostile conditions, Abdullah sang of his own responsibility, and even a sense of self-blame.[28] Perhaps this felt unavoidable; after all, no-one else seemed willing to help his community. But this was Abdullah's story, and, solidifying it in this way, he was also reclaiming some agency over it. In one last song, Abdullah emphasised this power, singing about *trying*: a term referring to attempts to reach the UK, often by climbing into lorries along the motorway.

I will try so much that people will be amazed
If the motorway is reached
I will try so much that people will be amazed
If the motorway is reached

If someone goes to Lidl
Or someone knows about Dunkirk
Immediately the smugglers will appear
The smuggler is a hajji, a commander, or a man with a moustache
When the motorway is reached
I will try so much that people will be amazed
If the motorway is reached

In Calais there are lots of Afghans, Black people, Kurds, and Iranians
On every rupee are ten smugglers
On ten rupees are sixty smugglers
If the motorway is reached
I will try so much that people will be amazed
If the motorway is reached

I will show you the way
And you yourself can try
Don't sleep too much
You should try by yourself
And don't sleep any more
You should try so that the smugglers' businesses collapse
If the motorway is reached
I will try so much that people will be amazed
If the motorway is reached

We are going with respect
It provides us with food:
May you live long! May you live long! Salam[29]
When it's two in the morning
Everyone is going to it
If the motorway is reached
I will try so much that people will be amazed
If the motorway is reached

Here, Abdullah is overtly resilient, defiant, active. Not only are he and his community going to defeat the money-hungry smugglers, but they're going to reach the motorway and defeat the border regime too.

PIECE
New Voice[30]

Think of the sound of your voice. What was the sound of your voice when you were born? What is the sound of your voice today? What does your voice express? How has your journey and the people you have met—and the people you have not met—affected your voice?

Resistance/Defeat

The Production of the Border

One evening, as I made my way out of the huge, makeshift camp, I was greeted by a small crowd, sitting on a grassy dune in front of the setting sun. The five young men had travelled from Ethiopia, all in their teens and early twenties. A brief 'hello' quickly and spontaneously turned into a chat about life and about music, until one of the men asked me if I had a pen and paper. He scribbled down some lyrics in the Oromo language, and explained, simply, that they were very important. Huddling over the paper to refresh their memories, the group transformed into an impromptu quintet, filling the evening air with a generous and beautiful chorus. Through some digging and searching back home, I eventually worked out that the words in my notepad were from a 90s song by Ebbisa Addunya, a musician and activist for Oromo liberation (Jalata, 2009). The largest ethnic group in Ethiopia, Oromo people have a long history of persecution (Jalata, 1998). A translated extract (Jalata, 2009) of the song reads:

I am proud of Gadaa[31]
I am the defender of Gadaa
If Oromos are united
Their enemies will be defeated

This was one of the more explicit instances of politicised music that I had witnessed in Calais. Forced from their homes in Ethiopia, and living destitute at European borders, these young people were propelled by a sense of hope, resilience, and conviction. And these seemed, to some degree, to be held in this song, supporting the group to see their convictions through, from Ethiopia to the Franco–British border.

Earlier, I considered how borders become expanses, dominated by certain people and certain, exceptional behaviours. At the centre of all this is the concept of the border itself, which, while it might function as an expanded space, is, after all, just the marking of a boundary. We often think of these markings as natural, as organic, as, simply, facts. But as Étienne Balibar (2004: 109) argues, 'natural borders' are a political myth. 'Everything here is historical, down to the linear character of borders as they appear on maps.' Borders are, simultaneously, a construction *by* the state and *of* the state: historical institutions whose juridical definition, political function, and the ways in which they are drawn, recognized, and crossed are liable to transform over time (Balibar, 2004: 108). While borders are, in principle, stable, fixed lines around the edge of a territory, in reality they are produced, changeable, and even appear in the middle of political space. (This is seen, for example, in ongoing events in Palestine,[32] at the Ethiopia–Sudan border,[33] and in the movable, stretching border spaces discussed earlier). But, as Balibar (2004: 109) outlines, the supposed stability of borders gives the state 'the possibility of controlling the movements and activities of citizens without [borders] themselves being subject to any control.' At borders, 'political participation gives way to the rule of police. They are the *absolutely nondemocratic* [...] condition *democratic institutions*. And it is as such that they are, most often, accepted, sanctified, and interiorized' (original emphasis). Through this acceptance, these false ideas of stability, naturalness, and absoluteness, borders are able to be used, and manipulated, to harmful effect.

Socio-political structures such as governance and possession are often organised around concepts of space and boundaries. Some see borders as a way of implementing certain rights, outlining, for example, the confines of self-determination and distributive justice (Banai, 2013). That is, borders show which people (those within a specified area) should have a say in the political, economic, social, and cultural systems of a space; and they determine who should receive a share of the material and social benefits and burdens (including wealth, health services, education). But borders also indicate who is *not* entitled to these rights. They allow people to differentiate between 'nationals' and 'foreigners,' and to define the 'worthy' from the 'migrant.' Supporting the notion that one group is more valid than the Other, the myth of the natural border promotes the idea of nationalism as a natural truth. That is, the unfounded and dangerous belief that if the border, and thus the nation, are natural, then the differences between people at each side of the border must also be natural, and accompanying nationalism thus a justified 'truth' (Lefebvre, 1991). Territories within borderlines 'are filled with normative content' (Robertson, 2009: 7), built on ideas of identity and belonging: components of nationality. In defining and encouraging national norms (even if this appears 'natural' or subtle), the state also defines what is abnormal, what is, in other words, an exception, and a threat to national order. Thus, the production of the border is inextricably linked to exception, to creating Others who don't fit into the space inside the border, and to creating the state and space of exception, as discussed above. Power, writes Mbembe (2003: 16), 'continuously refers and appeals to exception, emergency, and a fictionalized notion of the enemy. It also labors to produce that same exception, emergency, and fictionalized enemy.' In creating and upholding the myths that surround borders and those outside of them, states are also creating and upholding supposed justification *for* the border.

On the 24th of November 2021, in the English Channel, the owner of a fishing boat spotted several human bodies, floating next to a deflated dinghy. During the previous night, French and British emergency services had received numerous calls for help from the boat. Instead of sending out an immediate rescue, the countries spent hours passing responsibility (El Idrissi & Pascual, 2022; Taylor, 2022). Two people survived. Twenty-seven more bodies were taken from the water. Four were never found. Deaths at the Franco–British border happen so often that many in Calais know where to go for the following day's vigil. On the 30th of May 2022, I joined one such gathering in front of a town park. The previous day, in Beau-Marais, a suburb with small pockets of camps, Meretese Kahsay, had been hit by a train[34] (Observatory of Deaths at the Borders, n.d.; Marot, 2022). A huge banner was laid onto the ground, just big enough to hold the names of the 352 displaced people who had died at the Franco–British border since 1999. Traffic rumbled behind short speeches and an unaccompanied hymn, sung by one French mourner. Alongside this brief, melodic manifestation of grief and solidarity, the space felt notably still, quiet. It was a noticeable contrast to other gatherings of this displaced and non-displaced community, where sound—of activity, protest, music, or all three—was often at the forefront.

Later that evening I joined other members of the community in the form of the *Chorale de Lutte*, an open choir with a multilingual repertoire, based around border resistance. During a break, I spoke with some members, who told me about an incident from the days before. A small contingent of the group had been singing at a charity food distribution, hoping to provide some degree of entertainment and solidarity with those waiting in the long line. Responding to her signals, the singers placed themselves around a Sudanese woman, so she could be absorbed in the sounds and in this rare sonic moment. The police arrived. From the crowd they singled out the woman, and, ordering her to present identification papers, pulled her to the side, forcing her out of the moment and out of the music. Whether these officers were feeling particularly vindictive, whether they wanted to clamp down on any hint of pleasure, or whether they feared this sound-based communion might lead to a bigger, more active resistance, their response suggests that they recognised, however consciously, the potential power in music.

The young Oromo men I met on the grassy bank back in 2015 shared a song with explicit political, social, and personal meaning. It was a song from home, a song that united the group, and that came from a resistance movement that was relevant to their ongoing situation. But beyond its unambiguously political nature, their music appeared to carry something more holistically human, intrinsic, visceral. Despite, in this context, being attached to unjust and precarious conditions, music can provide feelings of security that may even be amplified in the context of exile. According to Alan Lomax (1959), as children acquire the language and emotional patterns of their culture, they also learn its musical style. He proposes,

> the primary effect of music is to give the listener a feeling of security, for it symbolizes the place where he was born, his earliest childhood satisfactions, [...] his pleasure in community doings [...] any or all of these personality-shaping experiences.
>
> *(p929)*

As the group sang and spoke about music, their emotional states were markedly transformed: from a quiet sadness in general conversation, to open, generous enthusiasm. For these men, music was clearly vital.

Of course, music isn't always a priority. When I asked one resident about his relationship to music, as the camp sound system played in the distant background, he replied 'Me, pffff ... Right now I've got many [other] problems.' Another time, someone asked: 'How can we have music here?

Only suffering.' These statements tell volumes about the harsh, wearing reality of displacement, borders, and the situation in Calais. Yet even within these extreme conditions, despite the associated inconvenience and demands on energy, music is meaningful and potent enough that it does persist.

Our behaviours may rarely, if ever, be completely conscious or intentional. According to Bourdieu (1977: 79) '[it] is because subjects, strictly speaking, do not know what they are doing that what they do has more meaning than they know.' This concept should be accompanied by a warning. In analysing and describing others' behaviours, we risk imposing our own ideas, retelling stories to prioritise our own expectations. At the border, where people are already dominated, disempowered, and transformed into the subjects of other people's systems and ideas, this risk is heightened. But Bourdieu also reminds us that politics and resistance exist on multiple levels. Not all the music I have discussed here takes place as a direct act of protest or dissent; but all can be seen, at least to some degree, as political. In the context of the border, and of overt institutional oppression and inequality, it is difficult *not* to recognise the political nature of events. The musical gathering in the following vignette provides an apt example of this. While the event felt strong and defiant, at least, from my own perspective, it was not necessarily a calculated act of political resistance. Nonetheless, to gather and make loud, expressive music, in a space where those involved are barely allowed to sleep, find food or, simply, exist, requires commitment, energy, discretion, and community solidarity. This, in turn, inherently requires resilience, resistance, and defiance.

In the chilly late spring of 2022, non-profit Mosaïc organised a jam session. I followed the coordinates I'd been given, well out of Calais town, along and then away from busy main roads, onto a sleepy cul-de-sac, down a muddy footpath, and across a usually-quiet-and-uneventful grassy expanse. People had laid down tarpaulin and carpet, and carried a generous assortment of percussion and guitars into the space. Before long, a sizeable group had gathered, sharing, between them, a minimal amount of music-making experience, and a huge amount of enthusiasm and community.

Coming together through cold and challenging conditions, in a secluded corner away from the significant police presence and associated risks, the gathering itself felt like an act of resistance. This was the community's space, and the music confirmed it. Sometimes it was a pulsing force of coordinated rhythms, led by different people, standing to introduce new ideas; often it was free and unconstrained, a feisty sonic experiment of collective independence. A few people spent time learning and playing with basic guitar chords; then, with a light-hearted glance, one picked a fork from the ground and took it to the strings, strumming with conviction and defiant[35] experimentation. As the evening developed, the drumming became strong and sure, protest-like, full of statement and intention. Built on resilience and human resolve, this safe moment was a rare opportunity for unrestricted expression and (relative) freedom. In this space within a border space, oppressive systems and restrictions were, perhaps, momentarily, at least symbolically, defeated.

PIECE
A Very Loud Silence

If an alarm sounds in the Mediterranean and no-one is available to listen, does it make a noise?

Outrospection/Introspection

Reflection and Reflexion

This chapter is unconventional in its form. My hope is that human stories—and your human responses—will lead, and that Pieces (to perform or reflect upon) will support a more comprehensive, individual exploration of the topic. Our own experiences shape our interpretations and understandings of events. Economically and politically dominant countries drive the situation I have discussed. This doesn't imply that other countries have no role—displacement, by its nature, concerns more than one place—but here I have focused on the Franco–British border. *I* am also a product of these places: a Brit who has lived either side of the Channel, and, conscious or not, the dominant patterns of thought in these places will have an impact on my own. This includes my understanding of music. No matter how much I study or spend time within different cultures, my emotional and cognitive biases draw from my own relationships with music, and the so-called 'Western musical traditions' that have been around me for most of my life. This chapter leads with human stories, which I hope allows readers to respond before diving deeper into context and analysis. As feeling, experiencing humans, with unique physical and emotional histories, we should consider our own roles in the stories that we tell and interpret.

Where there are people, there is music. Since the huge, sprawling camp was demolished in 2016, moments of active music-making are less likely to be stumbled upon at the Calais border. With smaller, more vulnerable, and less centralised camps, finding the space, safety, and resources has become more difficult. But music continues. Through small sound system parties, skilfully-made instruments, loud-but-discreet jam sessions, and many other instances I haven't discussed here; from guitar workshops at a charity day centre, to draining phone batteries to stream music from distant home.

Through the brutality of borders—within exceptional spaces of abuse and denied rights, within ambiguous yet nondemocratic spaces that move and follow the people within them, within spaces that belong to the whims and prejudices of certain people to the detriment of others—music persists. It would be a relief to believe that music alone is The Answer, to show that music is powerful enough to, all by itself, change the world. But the fetishisation of music, and the assumption of powers we might *like* to see, can be just another way of disempowering the humans behind it. Music is powerful, but it doesn't stand alone. We also need legal and political reform, abolition of abusive forces, of spaces of punishment and capitalist greed, and we need greater human empathy.

Music might not be a stand-alone answer, but it contributes. And, at the border, where people are deprived of basic resources, rights, and control over their own lives, music serves: as a means of creating safe space, escaping, gathering community, taking charge and assuming one's own value, entertaining, communicating, documenting and exploring experiences, marking events, sharing, reappropriating stories and reclaiming agency, affirming identity and human validity, making political statements, comforting and bringing joy, feeling secure, feeling empowered, owning the space, resisting.

Making music at the border isn't always convenient, but, somehow, people make it happen. Music is, after all, an intrinsic expression of being human, and one that most people can relate to. If we can recognise and hear these displaced music-makers as human, perhaps we can also recognise and hear them within the context of the border regime. And if we can understand the experiences that arise from the production of these border spaces, perhaps we can reconsider the border regime altogether.

PIECE
Life Piece

Take a walk from one location to another, in silence.
Listen to your surroundings as if they were a concert.
Listen to your own sounds, as if you were a player in that concert.
Do this regularly.

Notes

1 Travelling by bike is often the most affordable option, and a convenient way to move around Calais and its outskirts.
2 See Outrospection/Introspection.
3 A traditional score provides a written representation of a musical composition, typically using notation to instruct or guide musicians on how to perform a piece. Here, I use the term in a non-traditional sense, written and presented as 'Pieces' to enable and encourage active, experiential understanding.
4 See Prosser (Chapter One), who explores more of Oliveros' approach, and the relationships between listening and new 'ways of knowing.'

5 You may like to read this one through, and then close your eyes as you perform each section.
6 A stringed instrument in the lute family, common to some Islamic cultures.
7 Precise figures are difficult to obtain, with constant movement and many inhabitants reluctant to talk to authority figures.
8 *Les Compagnies républicaines de sécurité*, a branch of the French national police, with roles similar to British riot police.
9 This information comes from my first-hand observations, from accounts provided by people living at the camp, and from others supporting them.
10 I use 'sound system' to denote a mobile, speaker-generator set-up, and not the sound system culture that originates in Jamaican reggae. I explicitly reference this use of the sound system later in the chapter.
11 Tear gas, or CS gas, is a chemical weapon, and illegal in warfare (as per the international 1925 Geneva Protocol and 1993 Chemical Weapons Convention). Its effects are fierce and frightening. Tear gas causes intense pain in the eyes, nose, mouth, throat, and lungs. It can cause temporary blindness, a feeling of suffocation and acute anxiety. Heavy doses and intense exposure can cause lung damage and chemical burns (Hansard, 1970).
12 The Windrush scandal revealed how hundreds, and potentially thousands, of legal British residents had been denied legal rights and subject to Hostile Environment policies. Some were detained and even deported (Gentleman, 2017; 2018).
13 Le tribunal administratif de Lille
14 Inspired by Pauline Oliveros' 'Sonic Meditation: Variations on Listening' (2013: 190).
15 Abdullah's songs were recorded in audio form in Calais, in May 2015, with Abdullah's approval and consent to share. They were translated from Pashto by Tom Wide.
16 *The Jungle* was the name given to the big camp at the time. It is also often used to talk about smaller camps, before and after this one.
17 The Windrush Scandal provides an example of the limitations of this 'theory.' See endnote 12.
18 Its full title is: 'Regulation (EU) No 604/2013 of the European Parliament and of the Council of 26 June 2013: establishing the criteria and mechanisms for determining the Member State responsible for examining an application for international protection lodged in one of the Member States by a third-country national or a stateless person (recast).'
19 The UK was a Dublin Member State until January 2021, the end of the Brexit transition period.
20 It's worth acknowledging, however, that here, in this chapter, while I can quote Abdullah's voice and describe the surrounding context, these stories are being told through me, and not directly, or in their entirety, by Abdullah himself.
21 Another reference to the camp.
22 To honour their poetic nature, I am leaving full interpretation of these lyrics to readers.
23 The discount supermarket on the edge of Calais; people would sometimes walk there from the camp.
24 Implying that the fees are excessive.
25 Transliterated from the Pashto lyrics, this word may allude to a location within the sandy dunes of the camp.

26 This verse is full of irony, and Omid is being mocked.
27 Not his real name.
28 This is my own interpretation, though informed by discussions with Abdullah and his community, who appeared cornered by the impossibility of their situation. In English, Abdullah explained to me that these particular lyrics were 'advice for my people in Afghanistan. We're living in a tent, we [don't] have food [to] eat. Stay in Afghanistan and do your study—' At this point a friend interrupted: '*If* you don't have problem[s].' 'Yeah, *if* you don't have problem[s],' confirmed Abdullah. Abdullah himself had been forced, unwillingly, to leave his home, where his life was in direct and very real danger.
29 A non-profit organisation whose work includes distributing meals.
30 Inspired by Pauline Oliveros' 'Your Voice' (1971: xx).
31 Gadaa is the Oromo system of social structuring and a way of life (Jalata, 1998).
32 Here, Israeli forces have seized land, demolishing tens of thousands of Palestinian properties and displacing their former inhabitants to settle its own population (Amnesty International, 2017).
33 The disputed Al Fushqa region is claimed by both Sudan and Ethiopia (De Waal, 2021).
34 Details of this incident are limited. Reports indicate that Kahsay had been on the railway tracks, wrapped in a sleeping bag, when he was hit at around 6am (Ligue suisse des droits de l'Homme, 2023; Marot, 2022; Delattre, 2022).
35 This feels particularly defiant in a border space that punishes differences, expression of self-identity, and behaviours that step outside of norms, rules, and accepted practices.

References

Agamben, G. (2005) *State of Exception*. Translated by K. Attell. Chicago and London: University of Chicago Press.

Amnesty International (2017) *Israel's Occupation: 50 Years of Dispossession*. Available at: www.amnesty.org/en/latest/campaigns/2017/06/israel-occupation-50-years-of-dispossession. Accessed 20 June 2023.

Arendt, H. (1962) *The Origins of Totalitarianism*. Cleveland and New York: Meridian Books.

Balibar, É. (2004) *We, the People of Europe?* Translated by J. Swenson. NJ: Princeton University Press.

Banai, A. (2013) Political Self-Determination and Global Egalitarianism: Towards an Intermediate Position. *Social Theory and Practice*, 39(1): 45–69.

BBC News (2015) David Cameron: 'Swarm' of migrants crossing Mediterranean. *BBC*, 30 July, available at: www.bbc.co.uk/news/av/uk-politics-33714282. Accessed 15 Dec 2022.

Bourdieu, P. (1977) *Outline of a Theory of Practice*. Cambridge: Cambridge University Press.

Bradley, L. (2001) *Bass Culture*. London: Penguin Books.

Calais Migrant Solidarity (2015a) Police Fire Teargas into Calais Migrant Camp while People sleep. Available at: www.youtube.com/watch?v=fBy1wBBs8w8. Accessed: 4 Nov 2022.

Calais Migrant Solidarity (2015b) Violences Policières _ Jungle de Calais, Nov. Available at: www.youtube.com/watch?v=7R94JqCegpI. Accessed: 4 Nov 2022.

De Waal, A. (2021) Viewpoint: Why Ethiopia and Sudan have fallen out over al-Fashaga. *BBC*, 3 Jan, available at: www.bbc.co.uk/news/world-africa-55476831. Accessed: 20 June 2023.

Delattre, J. (2022) Calais: Un Migrant tué par un Train Alors Qu'il Dormait Sur La Voie Ferrée. *La Voix du Nord*, 29 May, available at: www.lavoixdunord.fr/1185987/article/2022-05-29/un-migrants-meurt-percute-par-un-train-de-marchandises-calais. Accessed: 20 July 2023.

DeNora, T. (2000) *Music in Everyday Life*. Cambridge: Cambridge University Press.

ECHR (2021) *"Dublin" Cases*. Available at: www.echr.coe.int/Documents/FS_Dublin_ENG.pdf. Accessed: 11 Dec 2022.

El Idrissi, A. & Pascual, J. (2022) Mort de 27 Migrants Dans la Manche: Les Enquêteurs Evoquent la « Non-assistance à Personne en Danger. *Le Monde*, 22 Nov, www.lemonde.fr/les-decodeurs/article/2022/11/21/mort-de-27-migrants-dans-la-manche-les-enqueteurs-evoquent-la-non-assistance-a-personne-en-danger_6150926_4355770.html. Accessed: 1 Dec 2022.

Equality and Human Rights Commission (2020) *Public Sector Equality Duty assessment of hostile environment policies*. Available at: www.equalityhumanrights.com/en/publication-download/public-sector-equality-duty-assessment-hostile-environment-policies. Accessed: 10 Nov 2022.

Gentleman, A. (2017) "I Can't Eat or Sleep": The Woman Threatened with Deportation after 50 years in Britain'. *Guardian*, 28 Nov, available at: www.theguardian.com/uk-news/2017/nov/28/i-cant-eat-or-sleep-the-grandmother-threatened-with-deportation-after-50-years-in-britain. Accessed: 29 April 2023.

Gentleman, A. (2018) Windrush Victims say Government Response is a "Shambles". *Guardian*, 8 Jun, available at: www.theguardian.com/uk-news/2018/jun/08/windrush-victims-say-government-response-is-a-shambles. Accessed: 29 April 2023.

Gentleman, A. (2020) Home Office Broke Equalities Law with Hostile Environment Measures. *Guardian*, 25 Nov, available at: www.theguardian.com/uk-news/2020/nov/25/home-office-broke-equalities-law-with-hostile-environment-measures. Accessed: 10 Nov 2022.

Gentleman, A. (2023) People Suspected of Living Illegally in UK to have Bank Accounts Closed. *Guardian*, 6 April, available at: www.theguardian.com/uk-news/2023/apr/06/people-suspected-of-living-illegally-in-uk-to-have-bank-accounts-closed. Accessed: 6 April 2023.

Gisti (1991) Roissy: Un Filtrage Sélectif. *Plein droit*, 13 March, available at: www.gisti.org/doc/plein-droit/13/roissy.html. Accessed: 9 Dec 2022.

Go, M. (2017) L'arrêté Municipal Visant à Interdire la Distribution de Repas aux Migrants Signé. *La Voix du Nord*. Available at: www.lavoixdunord.fr/126090/article/2017-03-02/l-arrete-municipal-visant-interdire-la-distribution-de-repas-aux-migrants-signe. Accessed: 1 Dec 2022.

Gower, M. (2022) *Irregular Migration: A Timeline of UK-French Co-operation*, Briefing Paper no. 9681. London: House of Commons Library. Available at https://researchbriefings.files.parliament.uk/documents/CBP-9681/CBP-9681.pdf. Accessed: 12 Dec 2022.

Grierson, J. (2018) Home Office Suspends Immigration Checks on UK Bank Accounts. *Guardian*, 17 May, available at: www.theguardian.com/uk-news/2018/may/17/home-office-suspends-immigration-checks-on-uk-bank-accounts. Accessed: 29 April 2023.

Hansard (1970) *Cs Gas: Volume 798*, London: House of Commons. Available at https://hansard.parliament.uk/Commons/1970-03-19/debates/260890c6-2cdc-4e71-97f4-795e4c2e3ca8/CsGas. Accessed: 29 April 2023.

Hansard (2022) *Western Jet Foil and Manston Asylum Processing Centres*: Volume 721, London: House of Commons. Available at: https://hansard.parliament.uk/commons/2022-10-31/debates/F189CA88-FDF3-4018-905C-1CC8A1B76E28/WesternJetFoilAndManstonAsylumProcessingCentres. Accessed: 15 Dec 2022.

Hebdige, D. (1979) [2002] *Subculture: The Maning of Style*. London and New York: Routledge.

Henriques, J.F. (2003) Sonic Dominance and the Reggae Sound System Session, in M. Bull. and L. Back, (eds.) *The Auditory Culture Reader*. Oxford: Berg, pp. 451–480.

HM Chief Inspector of Prisons (2020) *Report on unannounced inspections of the UK short-term holding facilities at France–UK Borders*. London: Her Majesty's Inspectorate of Prisons. Available at: www.justiceinspectorates.gov.uk/hmiprisons/wp-content/uploads/sites/4/2020/03/France-web-2019.pdf. Accessed: 11 Dec 2022.

Hicks, D. & Mallet, S. (2019) *Lande: The Calais 'Jungle' and Beyond*. Bristol: Bristol University Press.

Home Office (2022) UK–France Joint Statement: Enhancing Co-operation against Illegal Migration. Available at: www.gov.uk/government/publications/next-phase-in-partnership-to-tackle-illegal-migration-and-small-boat-arrivals/uk-france-joint-statement-enhancing-co-operation-against-illegal-migration. Accessed: 15 Dec 2022.

Home Office (2023) *Immigration System Statistics, year ending March 2023: Detention – Summary Tables*. Available at: www.gov.uk/government/statistical-data-sets/immigration-system-statistics-data-tables-year-ending-march-2023. Accessed: 14 July 2023.

Jalata, A. (1998) *Oromo Nationalism and the Ethiopian Discourse: The Search for Freedom and Democracy*. NJ: Third World Books.

Jalata, A. (2009) The Role of Revolutionary Oromo Artists in Building Oromumma: The Case of Usmayyoo Musa and Ebissa Addunya. *Oromo Studies Association Conference 2009*, Atlanta. Available at: http://gadaa.com/GadaaTube/1820/2011/08/31/usmayyoo-mussaa- music-marathon/. Accessed: 20 June 2015.

King, S. (2002) *Reggae, Rastafari, and the Rhetoric of Social Control*. Jackson: University Press of Mississippi.

Kirkup, J. & Winnett, R. (2012) Theresa May Interview: "We're going to give Illegal Migrants a really Hostile Reception". *Telegraph*, 25 May, available at: www.telegraph.co.uk/news/uknews/immigration/9291483/Theresa-May-interview-Were-going-to-give-illegal-migrants-a-really-hostile-reception.html. Accessed: 4 Nov 2022.

La Cimade (2022) *Audience du 20 Septembre 2022 au Tribunal Administratif de Lille: Interdictions des Distributions d'eau et de Repas à Calais*. Available at: www.lacimade.org/wp-content/uploads/2022/09/Audience-du-20-septembre-2022-au-Tribunal-Administratif-de-Lille-_-Interdictions-des-distribution-deau-et-de-repas-a%CC%80-Calais-POUR-RESEAUX.pdf. Accessed 12 Dec 2022.

LaBelle, B. (2018) *Sonic Agency*. London: Goldsmiths Press.

Le Tribunal administratif de Lille (2022) *Interdiction des Distributions de Denrées Alimentaires Dans le Centre-ville de Calais*. Available at: http://lille.tribunal-administratif.fr/Actualites/Communiques/Interdiction-des-distributions-de-denrees-alimentaires-dans-le-centre-ville-de-Calais. Accessed: 12 Dec 2022.

Lefebvre, H. (1991) *The Production of Space*. Translated by D. Nicholson-Smith. Oxford and Cambridge: Blackwell.

Ligue suisse des droits de l'Homme (2023) Forteresse Europe: 52'760 migrants morts en 30 ans. *Ligue suisse des droits de l'Homme*, 16 Jun 2023, available at: www.lsdh.ch/forteresse-europe-52760-morts-30-ans. Accessed: 20 July 2023.

Lomax, A. (1959) Folk Song Style. *American Anthropologist*, 61(6): 927–954.

Makaremi, C. (2009) Governing Borders in France: From Extraterritorial to Humanitarian Confinement. *Canadian Journal of Law and Society / Revue Canadienne Droit et Société*, 24(3): 411–432.

Marot, A. (2022) Un Migrant Décède dans un Choc Avec un Train de Fret à Calais. *France Bleu*, 29 May. www.francebleu.fr/infos/faits-divers-justice/un-migrant-decede-dans-un-choc-avec-un-train-a-calais-1653811057. Accessed: 20 July 2023.

Massey, D. (1993) Power-geometry and a Progressive Sense of Place, in J. Bird, B. Curtis, T. Putnam, G. Robertson, and L. Tickner. (eds.). *Mapping the Futures: local cultures, global change*. London and New York: Routledge. pp.60–70.

Mbembe, A. (2003) Necropolitics. Translated by L. Meintjes. *Public Culture*, 15(1): 11–40.

McGuinness, T. & Gower, M. (2018) *Immigration Detention in the UK: An Overview*, Briefing Paper no. 7294. London: House of Commons Library. Available at: https://researchbriefings.files.parliament.uk/documents/CBP-7294/CBP-7294.pdf. Accessed: 29 April 2023.

Ministère de l'intérieur (2022) *Maintien D'un étranger en Zone D'attente*. Available at: www.demarches.interieur.gouv.fr/particuliers/maintien-etranger-zone-attente. Accessed: 09 Dec 2022.

Observatory of Deaths at the Borders (n.d.) France: Belgium: UK. Available at: https://neocarto.github.io/calais/en/. Accessed: 15 Dec 2022.

Ohana, D. (2019) Carl Schmitt's Legal Fascism. *Politics, Religion & Ideology*, 20(3): 273–300.

Oliveros, P. (1971) *Sonic Meditations*. Urbana, IL: Smith Publications.

Oliveros, P. (2013) *Anthology of Text Scores*. Kingston, NY: Deep Listening Publications.

Rancière, J. (2001) 'Ten Theses on Politics'. Translated by R. Bowlby, & D. Panagia. *Theory & Event*, 5(3).

Refugee Rights Europe (2018) Refugees and Displaced People in Northern France: A Brief Timeline of the Human Rights Situation in the Calais Area. Available at: https://refugee-rights.eu/wp-content/uploads/2018/10/History-Of-Calais_Refugee-Rights-Europe.pdf. Accessed: 2 Dec 2022.

Robertson, S. L. (2009) "Spatialising" the *Sociology* of *Education*: *Stand*-points, *Entry*-points, *Vantage*-points. Bristol: Centre for Globalisation, Education and Societies, University of Bristol. Available at: https://susanleerobertson.files.wordpress.com/2009/10/2009-spatialising-soc-of-ed.pdf. Accessed: 14 Dec 2022.

Rudd, A. (2018) *Home Secretary statement on the Windrush Generation*. 23 April. Available at: www.gov.uk/government/speeches/home-secretary-statement-on-the-windrush-generation. Accessed: 29 April 2023.

Schlembach, R. (2016) A Political Movement is Rising from the Mud in Calais. *The Conversation*, 10 Feb, available at: https://theconversation.com/a-political-movement-is-rising-from-the-mud-in-calais-53758. Accessed: 15 Dec 2022.

Schmitt, C. (2005) *Political Theology: Four Chapters on the Concept of Sovereignty*. Translated by G. Schwab. Chicago and London: The University of Chicago Press.

Schneck, D. J. and Berger, D. S. (2006) *The Music Effect: Music Physiology and Clinical Applications*. London and Philadelphia: Jessica Kingsley Publishers.

Seba, A. (2015) Migrants de Calais: Un Camp « Toléré » Voit le Jour Autour du Centre Ferry. *La Voix du Nord*, 24 March, available at: www.lavoixdunord.fr/art/region/migrants-de-calais-un-camp-tolere-voit-le-jour-ia33b48581n2731759. Accessed: 18 June 2015.

Taylor, D. (2022) UK and French Coastguards 'Passed Buck' as 27 People Drowned in Channel. *Guardian*, 12 Nov, available at: www.theguardian.com/uk-news/2022/nov/12/uk-french-coastguards-passed-buck-people-drowned-channel. Accessed: 1 Dec 2022.

The Detention Forum (2022) *Short Term Holding Facilities in Northern France: A Policy Paper*, April. Available at: https://detentionforum.org.uk/wp-content/uploads/2022/05/STHF-policy-paper-April-22.pdf. Accessed: 11 Dec 2022.

Travis, A. (2017) UK Banks to Check 70m Bank Accounts in Search for Illegal Immigrants. *Guardian*, 21 Sep, available at: www.theguardian.com/uk-news/2017/sep/21/uk-banks-to-check-70m-bank-accounts-in-search-for-illegal-immigrants. Accessed: 19 April 2023.

Trump, D. (2018) [Twitter] 19 June. Available at: https://twitter.com/realDonaldTrump/status/1009071403918864385?s=20&t=35edBPqo5A0I4OFyEsIN5w. Accessed 15 Dec 2022.

UK Government (n.d.) Find an Immigration Removal Centre. Available at: www.gov.uk/immigration-removal-centre. Accessed: 29 April 2023.

Veal, M. (2007) *Dub: Soundscapes and Shattered Songs in Jamaican Reggae*. Connecticut: Wesleyan University Press.

ACKNOWLEDGMENTS

Our thanks go to the additional May 2022 Sonic Rebellions conference contributors who participated:

Tahera Aziz, Kareem 'Lowkey' Dennis, J Diaz, Julia Eckhardt, Lambros Fatsis, Carly Guest, Jishnu Guha-Majumdar, Markus Hetheier, Isac Ionuț, Caleb Madden, Edward Martin, Mershen Pillay, Rachel Seoighe, Luca Soudant, Nicholas Torretta, Rémy-Paulin Twahirwa, Rachel Wilson, Elizabeth Veldon, and Zettie Venter.

Students from the University of Brighton Digital Music and Sound Arts degree course provided an audio-visual installation created by Hattie Emmins, Ben Hampshire, Josie Hooper, Jolie McCallum, and Talyn Sandu. The event was part of the 2022 Brighton Fringe festival and supported by Centre of Applied Philosophy, Politics and Ethics research group at the University of Brighton. Thanks to the conference photographer, Kamal Badhey. We acknowledge and appreciate the support provided by technicians, catering, and cleaning staff at the University of Brighton, those who volunteered to chair sessions, and all those who attended. Our thanks to the Cowley Club community centre in Brighton for hosting our after-party and to the spoken word poets and musicians who performed, including the Lewes Spoaken Word collective, DJ Terror Wogan, DJ Juice Willis, and live set from Noah 'Cyphon' Thompson.

Additional acknowledgments to Mandeep Sidhu for her support and encouragement to the editor, and to Bella Tomsett for proofreading support and tolerating existential grammar ponderings.

Our huge gratitude to all the peer reviewers of this book who generously offered their time and feedback. Our reviewers represent a range of expertise and experience as academics, post-graduates, early career researchers, and artist-practitioners from around the world, including Canada, Brussels, New Zealand, Norway, South Africa, the UK, and the United States:

Melissa Avdeeff, Rayya Badran, Kieron 'Kaptin' Barrett, Ziad Bentahar, Richard Bramwell, J Diaz, Julia Eckhardt, Lambros Fatsis, Vivienne Francis, Malcolm James, Brandon LaBelle, Stephen Mallinder, Edward Martin, Laharee Mitra, Mershen Pillay, Mandeep Sidhu, Jaspal Singh, Tom Six, Elizabeth Veldon, Zettie Ventre, Clare Woodford, and Nimalan Yoganathan.

A special thank you to Caroline Robinson, a Sonic Rebel who long inspired the editor. Without her longstanding support this project would not have been possible.

INDEX

acousmatics 11–12, 112–14, 117–29
acousmêtre 113, 117–19, 122, 126
acoustic ecology 9, 16, 18
activism 6, 8, 84, 98
agency 2, 6–7, 98, 104, 128, 135, 139, 146, 153, 162–4, 166, 173
Algeria 6, 11–12, 132–50
Allo le système 145–6
alpha male 11, 91–2, 94, 97–9, 101–5
Andalusi 12, 140, 142–4
anthropology 16
anti-war 66, 75, 85
architecture 31, 33
asylum 154, 155
aural diversity 26–7
austerity policies 10, 45, 48–50, 54, 114–15, 122

balaclavas 10, 64, 66, 78–81; *see also* masks
Ball, Hugo 65, 67, 69, 74, 80–1
Berlant, Lauren 126–8
Berry, Siân 49
Black Lives Matter movement 1, 66
Blackness 64, 80–1, 84
Blitz Theatre Group 112, 116, 125
borders 1–2, 12, 92, 95, 134, 144, 153–4, 160–2, 168–9, 171–2; borderlines 160–1, 166, 169
Bouteflika, Abdelaziz 11, 132–3, 141–2, 145, 147
Butler, Judith 104

cabaret 68–9
Cabaret Voltaire 65–6
Calais, France 12, 152–3, 155–60, 162–3, 165–6, 168–9, 171–2
Cameron, David 49–50, 75, 163
capitalism/capitalist 30
Cavarero, Adriana 79–80
censorship 10, 65, 83–4, 135, 138–9, 145, 148
chaabi 141–2
Chekhov's First Play 112, 125–9
Chicago 54, 66, 75, 76
Chief Keef 53, 66, 75; *see also* Drill, rap
Chion, Michel 11, 113, 117–19, 122–3, 126
Cinemascope 11, 112–13, 116–27, 129
colonialism 6, 8, 17, 20, 132–3, 136–8, 142
communities 7–8, 46, 48, 65, 72–3, 75, 83, 85, 104–5, 138, 145, 155, 159, 162
Connor, Steven 123–4, 128
Conservative Party 5, 48–50, 58
corruption 132, 137, 146–7
Covid-19 1, 23, 25, 27, 34, 78
crime 10, 50, 64–5, 72–3, 76–8, 80, 82–4
criminalise 47, 64, 65, 77, 82–4
critical listening positionality *see* positionality

Dadaism 11, 64, 66–7, 69–71, 74–5, 78, 80–1, 83

Index

Darija Arabic 11, 133–4, 137, 142, 145
Dead Centre theatre company 112, 125–8
deadness 11, 113, 122, 125, 127–9
decolonising and/or decoloniality 5, 10, 19, 136
Deep Listening 17–18, 21–2, 24, 34, 154
detention centres 160–1
diaspora 133, 138, 142, 144–5
Digga D 71, 77, 84; *see also* Drill, rap
discourse 6, 46, 75, 82–3, 91–3, 96, 98, 100, 102–5, 113, 121, 134–5, 143–4, 146–8
discourse analysis 92, 105
discrimination 47, 65, 78, 91, 112, 138–9
dislocation 11, 67, 113, 116, 122
dissent 2, 98, 120, 133–5, 139, 142, 144–5, 147–8, 171
Dolar, Mladen 119, 121, 127
Drill 3, 10–11, 54, 58, 64–8, 70–85; *see also* UK Drill
Duchamp, Marcel 66–7
Duggan, Mark 76

East London 10, 45–9, 51, 53, 60, 67
epistemology 7–8
equality 98, 112, 136; *see also* inequality
embodiment 57, 121
Ethiopia 167–8
ethnographic 45, 54, 58, 86
ethnomusicology 16
European Debt Crisis 11, 112–13
European Union 115

Feld, Steven 7–8, 16
feminism 5, 11, 91–2, 96–99, 103–5
Fitna 132, 138–9, 144–5
football stadia 6, 12, 139–42, 144–5, 147–8
Foucault, Michel 8, 11, 92, 96–8, 105, 123
Franco-British border 6, 12, 153, 158, 160, 162–3, 168–9, 172
freedom 2, 73, 79, 85, 113, 121, 123, 139, 145, 148
Fresh & Fit podcast 11, 91–5, 97–8, 101, 105
Front de libération nationale (FLN) 132, 136–7, 146
Fusha (Modern Standard Arabic) 133–4

gangs 65, 72–3, 75–7, 82, 85
gentrification 10, 14–6, 21, 23, 27–35, 37–40
governance 11, 52, 116, 153, 169
Greece 113–15, 122–3
Grime 3, 10, 45–9, 67, 70–1; *see also* rap

hegemony 8, 11, 57, 84, 91, 96–8, 103, 105, 120
Hennings, Emmy 65, 74–5
Henriques, Julian 158–9
hip-hop 3, 48–9, 70, 75, 77, 85, 136, 138, 144–7
Hirak movement 11–12, 132–8, 141–8
homelessness 23, 55, 58, 60
Honig, Bonnie 103
hooks, bell 5
Hostile Environment policy 157–8, 160
human trafficking 11, 92, 94–5
'hungry listening' 10, 19–20

Indigenous music 19–20
industrial action 3–4
inequality 10, 47, 50, 59, 65, 82, 91, 113, 132, 137–8, 148, 153, 158, 171; *see also* equality
injustice 5, 8, 14–15, 33–5, 38–40, 47, 58, 133, 137, 153, 158, 160
intentionality 6, 26, 28, 78, 99, 115, 118, 120, 156, 158, 162, 166, 171

Jarvis, Jill 136

Kaba, Chris 1, 76
knife crime 64–5, 72–3, 75–6, 82–3; *see also* crime
Krept and Konan 71, 84

labour (party): New Labour 46, 49
LaBelle, Brandon 1, 7, 9, 104, 118, 128, 163
La Casa del Mouradia 141–4
language 5, 8, 10, 36–7, 54, 58, 60, 64, 67, 69, 73–5, 77, 79, 83–4, 91, 93, 96, 101–3, 119, 133–4, 137–8, 143, 146, 170; *see also* linguistics
Lazzarato, Maurizio 114–15
le pouvoir (Algeria) 132–3, 145–6, 148
'Levelling Up' agenda 45, 57–8, 60
liberty *see* freedom
linguistics 6, 11, 70, 75, 85, 101, 133, 138–9; *see also* language

listening walks 15, 22, 25, 27–8, 30, 32, 34–9
listening-with 10, 15, 21–2, 27, 35, 39–41
lived experience 2, 19, 45, 47, 48, 57, 58, 113, 147
Lounès, Matoub 138–9
lyrics 12, 48, 49, 51, 56, 57, 70–4, 77, 82–4, 86, 142, 145, 146, 174, 175

masculinity 11, 57, 96–7, 100, 105, 106
masks 1, 10, 64, 66, 78–81, 84; *see also* balaclavas
McKittrick, Katherine 5–6, 9
mediation 7, 128
memes 11, 71, 91–2, 97–105, 135, 147; counter-memes 97–8, 100, 102, 105
Metropolitan Police 75, 76, 86
Meziane, Raja 145–6
migrants 36, 101
misogyny 11, 91–9, 101–3, 105; *see also* sexism
Mosaïc 158, 171
Mouloudia Club d'Alger (MCA) 140–1
musicians 12, 52, 59, 70, 79, 135, 138–40, 143, 145–6, 180

narration 11, 31, 45, 48, 57–8, 60, 113, 116–19, 122–3, 125–9
narrative 5, 12, 58, 94, 99, 104, 119, 134, 137, 146
nationalism 100, 137, 143, 144, 169
neoliberal 5, 10, 11, 45–60, 98, 112–29
neoliberalism 3, 5, 8, 10–12, 49–50, 52–3, 59–60, 98, 112–16, 123, 125–9
New York 55, 58, 66, 71
non-literary forms 5, 92

Oliveros, Pauline 15–8, 20–1, 26, 34, 39, 154
oppression 2, 46, 112–16, 120, 128–9, 136, 158, 171
organic intellectuals 10, 45–6, 48, 60
Ouled el Bahdja 140–1, 148

parody/parodying 11, 68, 91, 92, 101, 102
participatory ethos 17, 22–3, 40
participatory methods/research 7–8, 10, 14–7, 20, 22–4, 27–8, 34, 40

patriarchy 11, 92, 96, 105, 143
podcasting 11, 91–9, 101–5
poetry 12, 65–6, 69–70, 74, 138–9, 154, 160, 164, 180
policing 8, 11, 12, 37, 38, 65, 73, 75, 76, 78, 83, 85, 98, 101, 139
positionality 14, 19–21, 26–7, 33–5, 40
poverty 31, 47–8, 50, 59, 84
public space 2, 9–10, 12, 33, 35–9, 50, 64, 85, 103, 139, 144, 147, 158
Pythagoras 114

racism 11, 36–7, 47, 81, 83, 85, 101, 138
rap 10, 12, 45–60, 64, 65, 67, 70, 71, 73, 76, 77, 79, 80, 82, 84, 85
ready-made 64, 67–71
redevelopment 15, 28, 32–3
reflective listening 10, 24, 33–5
reflexive 19–21, 27, 34, 39
refugee 12, 155, 163
Robinson, Dylan 10, 15, 17, 19–21, 34, 39

scores 5, 12, 154
sensory 8, 25, 40
sexism 91, 102–5; *see also* misogyny
social class 3, 10, 18–19, 33–4, 36–8, 45–7, 50–1, 60, 96, 104, 138, 140
social justice 2, 8, 12, 14–16, 20, 35, 39–40, 132, 136, 147–8
social media 2, 5–6, 9, 11, 23, 48, 56, 65, 71–2, 76–7, 82, 91–3, 95–7, 100, 103–5, 135, 142, 144–5
solidarity 6, 9, 12, 105, 139, 153, 157, 169–70
sonic agency 7, 9, 104, 163
sonic architecture 2, 10, 12, 91
'sonic vernacular' 135, 147
soundscapes 7, 16, 18, 30, 33–4, 37, 39–40, 134
sound system 12, 156, 158, 170, 172, 174
soundwalking 16, 17
Speier, Hans 101–3
structural violence 47, 50, 59–60
subversion 9–11, 75, 97, 100, 102–3, 125
suppression 2–3, 84–5, 138, 142
surveillance 46–7, 73, 77–9, 158

Tate, Andrew 94–5, 97, 105; *Tate Speech* podcast 11, 94, 105

Tausig, Ben 135, 147
tik-tok *see* social media
Torki, Sofiane Bakir 11, 133–5, 142, 144, 149
trend *see* social media
Tzara, Tristan 69, 74, 80

UK Drill (UKD) 64–5, 69–70, 72–8, 80, 82, 84–5; *see also* Drill
Union sportive de la médina d'Alger (USMA) 140–2, 145
urban practice 15–6, 23, 31–3, 35, 37, 39, 47, 85, 134, 140, 142, 148
urban seaside 15, 23, 31, 32, 39

ventriloquist and/or ventriloquism 113, 120–9
Verstraete, Pieter 119–20

violence 2, 10–2, 45–8, 57, 64–6, 69, 72–8, 80, 82–5, 94–5, 105, 113, 115–16, 132, 138–8, 141, 145, 165

war 2, 10, 11, 64, 66, 67, 69, 70, 72, 75–7, 80, 85, 96, 136, 138, 147, 149
Wiley 46, 47; *see also* Grime; rap

young people 10, 45–60, 65, 72, 84, 105, 139
youth club 10, 45–61
youth work 50–1
YouTube 51, 54, 71, 72, 91, 92, 95, 98, 129, 141–2, 145–7, 150

Žižek, Slavoj 116, 121, 126

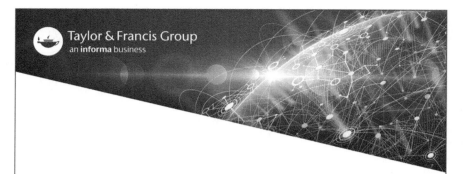